Patrick SCHWEITZER

帕特里克·施韦泽

（法）帕特里克·施韦泽　著　金明辉　译

国际建筑作品 2001—2021

Architectural International Projects & Works

2001 – 2021

辽宁科学技术出版社

·沈阳·

图书在版编目（CIP）数据

帕特里克·施韦泽国际建筑作品 2001—2021 /（法）帕特里克·施韦泽著；金明辉译 . — 沈阳 ：辽宁科学技术出版社，2022.7
ISBN 978-7-5591-2323-7

Ⅰ . ①帕… Ⅱ . ①帕… ②金… Ⅲ . ①建筑设计－作品集－法国－ 2001-2021 Ⅳ . ① TU206

中国版本图书馆 CIP 数据核字（2021）第 218186 号

出版发行：辽宁科学技术出版社
　　　　　（地址：沈阳市和平区十一纬路 25 号　邮编：110003）
印　刷　者：凸版艺彩（东莞）印刷有限公司
经　销　者：各地新华书店
幅面尺寸：245mm×290mm
印　　张：18.5
字　　数：370 千字
出版时间：2022 年 7 月第 1 版
印刷时间：2022 年 7 月第 1 次印刷
责任编辑：于峰飞
封面设计：周　洁
版式设计：周　洁
责任校对：韩欣桐

书　　号：ISBN 978-7-5591-2323-7
定　　价：328.00 元

联系电话：024-23280367
邮购热线：024-23284502
E-mail：1076152536@qq.com
http://www.lnkj.com.cn

目录 Contents

简介 Biography

帕特里克·施韦泽出生于1957年。1982年，他在导师迭戈·佩韦雷利的指导下顺利毕业，并成为一名建筑师。值得提到的一点是，建筑大师马里奥·博塔、奥雷利奥·卡菲提和路易吉·斯诺兹都曾担任其毕业答辩评委。此外，他还学习了社会学、民族学和艺术史等，1982年参加了法国财经部大楼竞赛。他的设计在克劳德·瓦斯克尼和让·努维尔等高手云集的竞赛中荣获了第6名的好成绩。1986年，他同欧蒂娜·戴克一同获得了"青年建筑师"称号。

帕特里克的第一批获奖作品包括霍因海姆体育馆（荣获建筑部门奖）和埃尔斯坦多功能厅（荣获安德烈·帕拉迪奥国际建筑奖）。他获得了"建筑记者"颁发的"聚光灯下的建筑师"马蒂厄·巴切洛特奖，在多米尼克·佩罗荣获这一荣誉的一年之后。1987年，帕特里克与皮埃尔·克内克特合作，随后设计了许多教育、体育以及文化类公共设施。他还参加了许多国内外竞赛，如位于马约特岛、留尼汪岛、马提尼克岛和圭亚那的项目。他设计了位于卡宴的圭亚那地区总部。他参加了新欧洲议会的竞赛（一等奖是AS建筑工作室）。

2001年，他创立了帕特里克·施韦泽联合建筑事务所。2008年，他操刀设计的第一个住宅项目斯特拉斯堡弗洛尔花园住宅荣获了维尔梅尔金字塔可持续住宅奖，这是阿尔萨斯地区第一座通过认证的低能耗住宅建筑。2010年，他与阿尔萨斯建筑设计院合作，参与了上海世博会阿尔萨斯地区馆（可再生能源示范建筑）的设计与建造。随后，他们共同参加了中国的众多建筑和城市规划竞赛项目。2011年，他设计的全木结构的米雷库尔农林业学校获得国家波巴特木材奖。3个住宅项目获得维尔梅尔金字塔银奖。

2013年，他在巴黎建筑美术馆举办个人作品展。2018年，他设计的基加利建筑与环境设计学院获得美国A+Architizer奖、亚太设计中心国际设计大奖、建筑大师奖荣誉奖以及德国设计大奖特别奖，此外，还入围了AFEX大奖和AMO建筑奖最具创意奖的决选。

Biography

Patrick Schweitzer was born in 1957. He graduated as an architect in 1982 under the guidance of Diego Peverelli. Mario Botta, Aurélio Galfetti and Luigi Snozzi were on his examination board. After studying sociology, ethnology and art history, in 1982 he took part in the competition for the Ministry of Economy and Finance in Bercy, where his project was ranked sixth, between the projects of Claude Vasconi and Jean Nouvel. It was published in February 1983 in Architecture d'Aujourd'hui. In 1986 he won the Albums de la Jeune Architecture, the same year as Odile Decq.

His first projects won awards, such as the sports hall in Hoenheim, which won the Departmental Architecture Prize, and the multi-purpose hall in Erstein, which won the Andréa Palladio International Architecture Prize. He also won the Mathieu Bachelot Trophy from the "journalistes de la construction", one year after Dominique Perrault. In 1987 he partnered with Pierre Knecht. He then designed numerous public facilities in the fields of education, sport and culture. He took part in numerous national and international competitions, including projects in Mayotte, Reunion, Martinique and Guyana. He designed the headquarters of the Guyana Region in Cayenne. He also participated in the competition for the new European Parliament won by Architecture Studio.

In 2001, he created his own agency Patrick Schweitzer & Associes. In 2008, he was awarded the Pyramide de Vermeil for sustainable housing for the «Les Jardins de Flore» project in Strasbourg, his first housing project and the first certified low-energy building in Alsace. In 2010, he took part in the creation of the Alsace Region Pavilion at the Shanghai World Expo, an exemplary building in terms of renewable energy, as well as in numerous architectural and urban planning competitions in China, in partnership with AADI Alsatian Architectural Design Institute. In 2011, he received the national Bobat Wood Award for the regional agricultural high school in Mirecourt, an entirely wood-framed building. Three silver pyramids were awarded for housing projects.

In 2013 he exhibited at the Architecture Gallery in Paris. In 2018, the project for the Kigali School of Architecture in Rwanda won three architectural prizes in the USA, the A+Architizer Award, the *IDA APDC prize and a mention at the Architecture Masterprize, as well as a special mention at the German Design Award. He is also a finalist of the Grand Prix of the AFEX Architectes Français à l'Export and of the AMO Architectes et Maîtres de l'Ouvrage.

奖项 Awards

2021年
—奥伯罗斯伯尔让的冥想室荣获2021年大东部地区木结构建筑奖室内设计奖

2020年
—下莱茵省档案馆住宅改造荣获2020年建筑社区奖

—苏弗尔韦埃尔桑生态住宅区荣获2020年建筑社区奖二等奖

—嘉士伯欧洲研发中心荣获 2020 年建筑社区奖荣誉奖

—斯特拉斯堡影像引导外科研究所荣获2020年建筑社区奖荣誉奖

—Urban Side住宅荣获2020年建筑社区奖三等奖

—埃尔斯坦多媒体图书馆荣获2020年建筑社区奖三等奖

—新诺尔杜斯市政厅荣获2020年建筑社区奖荣誉奖

—基加利建筑与环境设计学院荣获2020年A设计大奖赛金奖

—Quiri棕地住宅改造荣获2020年重新思考未来奖（三等奖）

—下莱茵省档案馆住宅改造荣获2020年重新思考未来奖（一等奖）

—下莱茵省档案馆住宅改造荣获德国设计大奖卓越建筑奖

—Quiri棕地住宅改造荣获德国设计大奖卓越建筑奖

2019年
—荣获2019年BUILD建筑奖"法国最佳医疗建筑公司"奖

—基加利建筑与环境设计学院荣获2019年城市设计建筑奖一等奖

—Quiri棕地住宅改造荣获2019年城市建筑设计奖三等奖

—埃尔斯坦精神病院荣获2019年城市设计建筑奖三等奖

—下莱茵省档案馆住宅改造荣获2019年城市设计建筑奖荣誉奖

—基加利建筑与环境设计学院荣获2019年杜伊弗建筑奖评委会荣誉奖

—埃尔斯坦精神病院荣获2019年建筑大师奖荣誉奖

—下莱茵省档案馆住宅改造荣获2019年建筑大师奖荣誉奖

—荣获2019年总承包商金钥匙奖及国家可持续发展创新奖提名

—基加利建筑与环境设计学院荣获2019年标志性设计奖创新建筑奖

—基加利建筑与环境设计学院荣获2019年全球未来设计大奖（一等奖）

—下莱茵省档案馆住宅改造荣获2019年全球未来设计大奖（二等奖）

—Quiri棕地住宅改造荣获2019年全球未来设计大奖（三等奖）

—Quiri棕地住宅改造荣获2019年ENSAS学生颁发的金字塔银奖

—基加利建筑与环境设计学院荣获Archilist"月度项目"奖

—Quiri棕地住宅改造荣获2019年设计灵感奖玫瑰金奖

—下莱茵省档案馆住宅改造荣获设计灵感奖银奖

2018年

　　—基加利建筑与环境设计学院入围2018年AMO建筑奖最具创意奖

　　—基加利建筑与环境设计学院荣获2018年建筑大师奖荣誉奖

　　—基加利建筑与环境设计学院荣获2018年德国设计大奖特别奖

　　—基加利建筑与环境设计学院荣获2018年亚太设计中心国际设计大奖

　　— 基加利建筑与环境设计学院荣获2018年美国A+Architizer奖

　　—基加利建筑与环境设计学院入围2018 年AFEX大奖

2016年

　　—斯特拉斯堡城郊工厂住宅荣获下莱茵省委员会特别奖（金字塔银奖）

2011年

　　—斯特拉斯堡Opus花园住宅荣获2011年天蓝色可持续发展奖（金字塔银奖）

　　—斯特拉斯堡生态极之门住宅荣获2011年GDF天然气奖（金字塔银奖）

　　—米雷库尔农林业学校荣获2011年国家波巴特木材奖

2008年

　　—荣获国家金字塔金奖

　　—斯特拉斯堡弗洛尔花园住宅荣获2008年维尔梅尔金字塔可持续住宅奖

　　—斯特拉斯堡弗洛尔花园住宅荣获2008年地区大奖（金字塔银奖）

　　—斯特拉斯堡弗洛尔花园住宅荣获2008年可持续住宅奖（金字塔银奖）

　　—斯特拉斯堡弗洛尔花园住宅荣获2008年天蓝色奖（金字塔银奖）

　　—斯特拉斯堡弗洛尔花园住宅荣获2008年地产设计奖（金字塔银奖）

　　—斯特拉斯堡弗洛尔花园住宅荣获2008年商业地产奖（金字塔银奖）

1990年

　　—霍因海姆体育馆荣获建筑部门奖

1989年

　　—埃尔斯坦多功能厅荣获安德烈·帕拉迪奥国际建筑奖

1987年

　　—荣获"聚光灯下的建筑师"马蒂厄·巴切洛特奖

1986年

　　—荣获"青年建筑师"称号

Awards

2021

- Winner of the Prix Régional de la Construction Bois Grand Est 2021 in the category of interior design for the meditation room in Oberhausbergen

2020

- Winner The Architecture Community 2020 for the former Archives of the Bas-Rhin
- Second Award The Architecture Community 2020 for the eco-neighbourhood in Souffelweyersheim
- Honourable Mention The Architecture Community 2020 for the Carlsberg Research Centre
- Honourable Mention The Architecture Community 2020 for the IHU in Strasbourg
- Third Award Mention The Architecture Community 2020 for Urban Side
- Third Award Mention The Architecture Community 2020 for the Erstein media library
- Honorable Mention The Architecture Community 2020 for the new Nordhouse Town Hall
- Gold A'Design Award 2020 for the Kigali School of Architecture
- Rethinking The Future Awards 2020 for the rehabilitation of the Quiri industrial wasteland into housing (3rd prize)
- Rethinking The Future Awards 2020 for the rehabilitation of the former departmental archives of the Bas-Rhin into housing (1st prize)
- Excellent Architecture Award at the German Design Awards for the renovation of the former departmental archives of the Bas-Rhin into housing
- Winner of the German Design Awards Excellent Architecture for the rehabilitation of the Quiri industrial wasteland into housing

2019

- Best Healthcare Architecture Studio 2019 - France», BUILD Architecture Awards 2019
- Winner of the Urban Design Architecture Design Awards 2019 for the Faculty of Architecture and Environmental Design in Kigali
- Third Award of the Urban Design Architecture Design Awards 2019 for the rehabilitation of the Quiri brownfield site into housing
- Third Award of the Urban Design Architecture Design Awards 2019 for the psychiatric hospital in Erstein
- Honourable Mention of the Urban Design Architecture Design Awards 2019 for the rehabilitation of the former departmental archives of the Bas-Rhin into housing
- Honorable Mention and congratulations from the jury of the Duyver Prize 2019 for the Faculty of Architecture and Environmental Design in Kigali
- Honourable Mention Architecture MasterPrize 2019 for the Psychiatric Hospital in Erstein
- Honourable Mention Architecture MasterPrize 2019 for the rehabilitation of the former departmental archives of the Bas-Rhin into housing
- Nominated for the Clés d'Or 2019 de l'Entreprise Générale competition and the national innovation prize for Vert en Vue
- Innovative Architecture, Iconic Awards 2019 for the Faculty of Architecture and Environmental Design in Kigali
- Global Future Design Award 2019 for the Faculty of Architecture and Environmental Design in Kigali (1st prize)
- Global Future Design Awards 2019 for the renovation of the Archives Départementales du Bas-Rhin into housing (2nd prize)
- Global Future Design Awards 2019 for the rehabilitation of

the former Quiri wasteland into housing (3rd prize)
- Silver Pyramid Award given by the students of ENSAS, for the rehabilitation of the former Quiri wasteland into housing
- Archilist «PROJECT OF THE MONTH» Award, for the Faculty of Architecture and Environmental Design in Kigali
- Rose Gold» Muse Awards for the rehabilitation of the Quiri brownfield site into housing
- Silver» Muse Awards for the rehabilitation of the former departmental archives of the Bas-Rhin into housing

2018
- Finalist, AMO Most Creative Typology Award 2018, for the Faculty of Architecture and Environmental Design in Kigali
- Honourable Mention, Architecture Masterprize 2018, for the Faculty of Architecture and Environmental Design in Kigali
- Special Mention, German Design Awards 2018, for the Faculty of Architecture and Environmental Design in Kigali
- Winner, *IDA APDC Award 2018, for the Faculty of Architecture and Environmental Design in Kigali
- Winner, A + Architizer Awards 2018, for the Faculty of Architecture and Environmental Design in Kigali
- Finalist, AFEX Grand Prix 2018, for the Faculty of Architecture and Environmental Design in Kigali

2016
- Special prize from the Conseil Départemental du Bas-Rhin (Silver Pyramid) for the Urban Side Factory project in Strasbourg

2011
- Bleu ciel sustainable development prize 2011 (Silver Pyramid) for the Opus Garden project in Strasbourg
- GDF Natural Gas Award 2011 (Silver Pyramid) for the project

Les portes de l'Ecopole in Strasbourg
- Bobat Wood Award 2011 for the project of the regional agricultural high school in Mirecourt

2008
- Distinction in the national Golden Pyramid Awards 2008
- Vermeil Pyramid for the sustainable housing prize for the Les jardins de Flore project in Strasbourg
- Regional Grand Prize 2008 (Silver Pyramid) for the project Les jardins de Flore in Strasbourg
- 2008 Sustainable Housing Award (Silver Pyramid) for the project Les jardins de Flore in Strasbourg
- 2008 Bleu ciel Award (Silver Pyramid) for the project Les jardins de Flore in Strasbourg
- 2008 Property Design Award (Silver Pyramid) for the Les jardins de Flore project in Strasbourg
- 2008 Commercial Property Award (Silver Pyramid) for the Les jardins de Flore project in Strasbourg

1990
- Departmental architecture prize for the sports hall in Hoenheim

1989
- André Palladio International Architecture Prize for the multi-purpose hall in Erstein

1987
- Winner of the Architects in the Spotlight Mathieu Bachelot Prize

1986
- Winner of the Albums de la jeune architecture

序言 Foreword

　　理查德·韦斯顿在《材料、形式与建筑》一书中提到其海牙办公室附近由MVRDV事务所打造的住宅区项目时,写道:"这些房子有着相似的轮廓,多样的色彩以及令人出乎意料的物质属性。"[1]这个项目被称作世纪之交的象征,和其他建筑一起表达了现代主义基本原则的时代的终结以及追求理性主义的时代的开始。尽管如此,建于21世纪初的建筑仍被称为"超现代主义建筑",这是汉斯·伊贝林斯于1988年提出的一个术语。随后,人类学家马克·奥吉将其演绎为"超级现代性"。荷兰评论家曾经分析说:"在后现代和解构主义建筑之后,一种新型的建筑形式正在兴起,着重强调触觉、视觉和空间感[2]。"这意味着,建筑不再单纯突出结构的真实性,更不用说框架的重要性了。建筑俨然更关乎质地、密度、造型和开口等细节问题。从这一角度看来,建筑就更加简单了,"超现代主义"可以被视作后现代主义的最好成果,既摆脱了现代道德的束缚,又摒弃了早期的过度批判。

　　浏览帕特里克·施韦泽联合建筑事务所的作品就仿佛是在欣赏建筑史上这一特定的时刻——建筑行业正面临着多重矛盾和挑战的时刻。当代建筑集合理性、物质性和对背景的敏感度于一身,是现代主义遗产的一部分,但其必须要适应当下的环境并能够以全新的敏感性和崭新的质地来丰富现代主义传统。总体而然,其植根于资源经济——这是前卫派的共同理念,不像20世纪的建筑那么激进,重要的是对现有形式和材料的保护和再利用。帕特里克·施韦泽拥有20多年的从业经验,

[1] 韦斯顿. 材料、形式与建筑[M]. 巴黎:1e Seuil出版社,2003.
[2] 伊贝林斯. 超现代主义:全球化时代的建筑[M]. 巴黎:Hazan出版社,2003:132.

在每一个项目的设计中找到了共通点,如今却面临着来自理论和道德方面的多重挑战:不要抵制作为文化遗产的现代运动;充分考虑当地特色并保持一定的警惕性和批判性;将必要性转化成优势;充分融入周围的环境中;打造良性的建筑。自然而然地接受曾经存在的建筑,没有必要通过冗长的演讲来证明形式、材料和色彩的合理性。在建筑领域,简约是非常必要的,这与其他领域可能不大相同。对结构和环境而言,关联性和准确度是至关重要的。

帕特里克·施韦泽拥有社会学背景,从关爱的角度出发,处理了一些常见的与社会相关的主题。他曾设计过数个优秀的医疗建筑,无论在技术上还是形式上,都成了该领域的范例。这些专业知识不仅在法国传播,而且在非洲国家也得到广泛的认可。帕特里克·施韦泽于2017年在卢旺达设计了基加利建筑与环境设计学院,这使他与当地社会建立了牢固而持久的合作关系。建筑特有的红色混凝土棱柱造型与当地的林地环境完美融合,那些用于描述建筑、风格或隶属关系的术语在这里已经不再具备任何意义。现在,让我们继续回到"现代性"的主题上,它一直在影响着我们对建筑和艺术的思考。从更广泛的意义以及人类与居住环境之间的关系上来看,奥古斯丁·伯克一直坚信"现代性让世界脱节了"。这其实意味着现代性中存在倾向于破坏既定平衡的原则,或是整个社会的失衡,或是生态系统的失衡。现代性具有所谓的破坏性的同时,也具备更强大的改造事物的手段。这种双重作用的结果加剧了人类

和事物之间的分离感,从而导致我们生活的世界越来越不稳定[3]。在基加利项目中,并没有出现这种分离感,或者说它以一种合理的逻辑融入设计中。在诺尔杜斯或米雷库尔项目中,他们运用了同样的理念,创造出了可以称之为"中庸之道"的建筑。

西蒙·特赛尔

艺术与建筑历史学家

亚眠大学现代艺术史教授

[3] 伯克.生态人的伦理原则:地球上的人类[M].巴黎:Gallimard (Le Débat)出版社,1996:19.

Foreword

In the introduction of his book *Materials, Form, and architecture*, Richard Weston refers to the housing estate created by the office MVRDV near The Hague: the houses' familiar silhouettes; their diverse colours; their unexpected materiality[1] . Emblematic of the turn of the century, this project, like many others, expresses a liberation from the basic principles of modernism, starting with constructive rationalism. The architecture of the beginning of the 21st century has nonetheless been interpreted as belonging to supermodernism, a term suggested in 1998 by Hans Ibelings. This was shortly after the anthropologist Marc Augé described the figures of excess which, according to him, characterised supermodernity. After postmodern and deconstructivist architecture, which mainly called upon the intellect, an architecture is arising that privileges touch, visual and spatial sensations[2], analyses the Dutch critic. It is no longer a matter of the truth of the programme, and even less of highlighting the framework. It has become a matter of textures, density, unconventional forms and random openings. From this perspective, which is indeed simplistic, supermodernism is best conceived as an accomplishment of post-modernism, as it is freed from both modern morals as well as the excesses of its early criticism.

To go through the works of Schweitzer & Associés Architectes is to appreciate this specific moment in the history of architecture, where a profession is being confronted with multiple and sometimes contradictory challenges. Whilst assuming a part of the modernist

1 Richard Weston, *Formes et matériaux dans l'architecture* (Translated from English by Pierre Saint-Jean), Paris, le Seuil, 2003.
2 Hans Ibelings, *Supermodernisme : l'architecture à l'ère de la globalisation*, Paris, Hazan, 2003 (1ère éd. Amsterdam, 1998), p. 132.

legacy which it must adapt to environmental requirements as well as enriching this modernist tradition with new sensibilities and new textures, today's architecture combines rationality, materiality and sensitivity towards context.

Overall less radical than the architecture of the previous century, it is rooted in the economy of resources - a notion shared with the avant-gardes - but above all in the conservation and reutilisation of existing forms and materials. In twenty years of projects where empathy can be found in every design, Patrick Schweitzer has faced this theoretical and ethical challenge: not to reject the Modern Movement as a cultural heritage, to be vigilant and critical by considering the local dimension, to finally transform necessity into virtue, to be humble enough to blend into the surrounding landscape, into old walls, and to design virtuous architecture. What was once stance is now nature; there is no need, therefore, for long speeches to justify choices of form, materials or colour. Simplicity is required here, fantasy elsewhere. Everything is a matter of relevance, of accuracy in terms of programme and environment.

With a background in sociology, Patrick Schweitzer has also tackled a number of themes that shape our society, starting with the notion of care. Several outstanding medical projects, in both technical and formal terms, have made the studio a reference in this field. With both discretion and confidence, this expertise is disseminated in France and now also in Africa. A strong and lasting relationship has been established with Rwanda in particular, where Patrick Schweitzer made his debut in 2017 with an exceptional project: the Faculty of Architecture in Kigali, whose red-painted concrete prisms fit seamlessly into a woodland environment. The terminologies commonly used to describe architecture, to determine its style or affiliation, are no longer of any meaning here. Let us return to the subject of modernity, which continues to haunt our thoughts about art and architecture. Looking at the latter in its broadest sense and questioning the relationship of humanity with its living environment - the ecumene - Augustin Berque argued that modernity disjoins the world. This means that in modernity there exist principles that tend to undo the established balances, whether these are the balances of societies or of ecosystems. Modernity constantly tends to drain symbols of their power of inclusion, at the same time as it offers ever more powerful means of transforming things; the result of this double motion being an increasing disjunction between things and human sensibility, and thus an increasing instability of what makes up the world we live in[3]. In Kigali, it is anything but disjunction that is at work; dissociation, even distinction, certainly infuses the project, but precisely within a logic of integration. In Nordhouse or Mirecourt, the same integrative thinking also guides the project's construction and creates an architecture that we could say is of the golden mean.

Simon Texier
Historian of Art and Architecture
Professor of the History of Contemporary Art
at the University of Picardie Jules Verne

3 Augustin Berque, *Êtres humains sur terre. Principes d'éthique de l'écoumène*, Paris, Gallimard (Le Débat), 1996, p. 19.

市镇公共社区
Community of Communes

2005年
莫尔塞姆 / 法国
Molsheim / France

莫尔塞姆是一座极具活力的工业城市,这里是世界名车布加迪总部和奔驰工厂等总部和工厂的聚集地。

米齐格区议会新总部选址于此,旨在为以前位于市政厅的机构提供现代且实用的场所,同时又能反映当地的蓬勃生机。设计的目标之一即能够让建筑顺应时代的不断变化。整栋建筑沿南北轴线展开,专门设置的中心结构既可以更好地服务于南北两侧翼楼,又便于未来东侧的扩建。建筑风格沉稳而现代,完美实现了建筑语言与法律法规的融合。主入口退于街道之后,长檐突出的门廊以及整齐划一的装饰彰显了建筑本身的功能。

整栋建筑共为两层,一层设置接待大厅、全体议会会议室和办公室,二层包含正、副区长办公室以及服务办公区。此外,还建有地下停车场、档案室及技术设备区。

值得提到的一点是建筑师特意打造的圆形全体议会会议室,打破了建筑整体方正规整的造型,内外皆可清晰看到。

之所以选择圆形是为了突出空间的功能——咨询与沟通,借以消除等级观念,传递出"无人拥有特权"的理念。

此外,会议室外观采用宽大的玻璃幕墙打造,与别处的孚日山脉砂岩外墙相得益彰。

The city of Molsheim is known worldwide for its industrial dynamism, hosting the headquarters and factories of major groups such as the car manufacturers Bugatti and Mercedez-Benz.
The new headquarters of the Molsheim-Mutzig district council is intended to provide contemporary and functional premises for the institution which was previously housed in the town hall, and to reflect the dynamism of the local area. The expectations for the new building include its ability to evolve over time. The building is structured along a north-south axis and organized around a central core that serves the two wings and anticipates a future extension to the east. Sober and contemporary, the used architectural language plays with the codes of institutional architecture. The setback from the street, the central entrance highlighted by a long canopy, the regularity of the ornamental elements and the row of large trees immediately suggest the official function of the building.

The program is spread over two levels; the ground floor includes the reception hall, the plenary and committee rooms, and offices. On the first floor are situated the offices of the district president and vice-presidents, as well as service offices. The building also includes a basement with parking, archives and technical rooms.
The clean orthogonality of the project is deliberately interrupted by the unexpected volume of the plenary room. Perfectly circular and protruding from the square it is set in, this curbed volume is visible from the inside as well as the outside.
Choosing a circular composition establishes the function of the place – consultation – and erases the notion of hierarchy: no one holds a privileged position.
The plenary room is generously glazed, allowing full visibility from the outside on the meetings taking place in the heart of the building which is clad with Vosges sandstone.

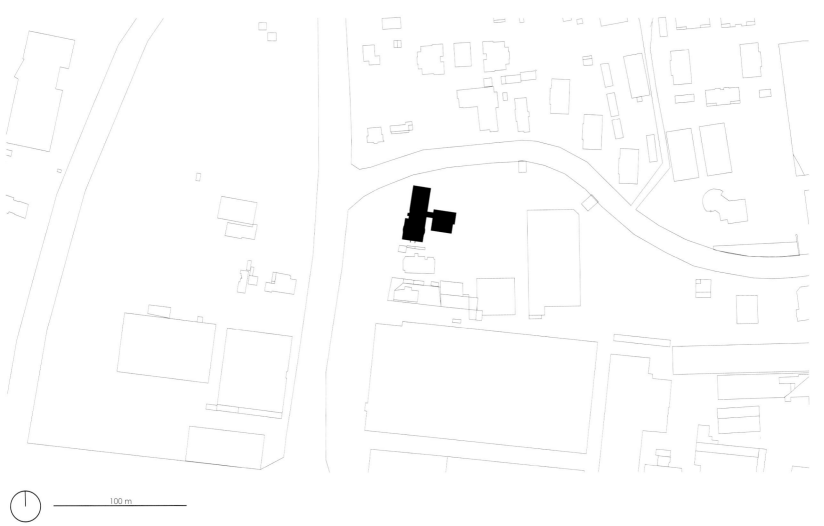

100 m

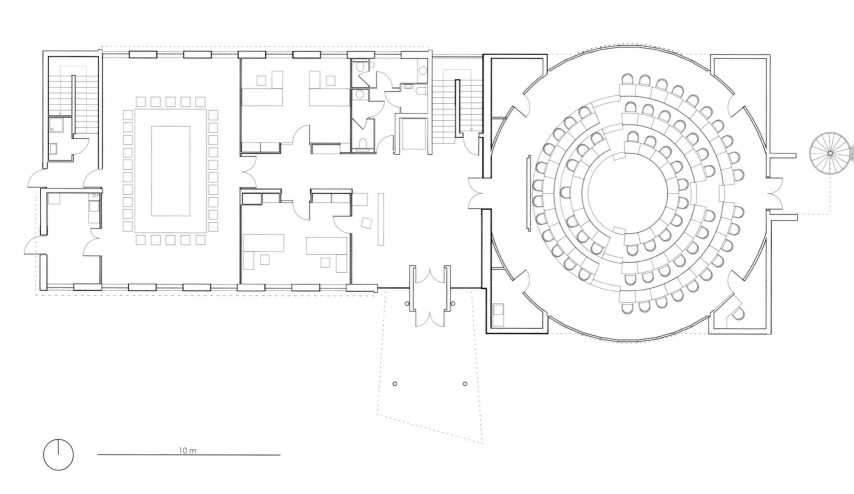

10 m

办公和商业
Office and Commerce

2008年
斯特拉斯堡 / 法国
Strasbourg / France

RMT改造项目是诺伊霍夫大道城市更新计划的一部分,包括商店、斯特拉斯堡保险公司总部、企业孵化区以及出租写字间等。

整栋建筑共为6层,带有地下室和阁楼,与城市景观和谐共融。建筑造型简约却不失时尚,既具备公共空间特色,又为凸显社区特征做出了杰出的贡献。

底层采用大面积玻璃结构打造,内部有各种商店,标识统一显示在建筑立面上。独特的梁柱系统赋予立面最大程度的自由,可根据需求进行单独处理,并直接呈现每家店面的不同特色。

建筑在诺伊霍夫大道一侧,因高大现代的外观而格外引人注目。外墙采用染色混凝土装饰,并点缀深色木隔板,进一步凸显了建筑本身的垂直度以及醒目的城市特征。

西立面的设计恰好与之相反,更突出人性化理念,强调室内外空间的过渡,如步入式屋顶露台及天井的设置。整栋建筑与相邻的公寓楼相互呼应,给人以十足的亲切感。

郁郁葱葱的绿色植物赋予建筑独特的体验——不可进入的屋顶上有当地耐寒植物,一楼布置了绿地和树篱,与大道上茂盛的大树交相呼应。

建筑材料以天然简单为主,如彩色混凝土、天然木材饰面以及铝制饰边等。这恰好说明了一个问题,即采用可持续的当地材料可以打造出现代时尚的建筑。

The RMT block is a service facility that includes shops, the headquarters of a Strasbourg insurance company, a business incubator and standardized offices for rent. This program is part of an urban area undergoing complete renewal, along the Avenue de Neuhof.
The project, a six-story building with a basement and an attic, is rooted in its harmonious integration into the urban landscape that frames it. Its sleek and humble architecture aims to shape the public space and to contribute to the neighbourhood's identity.
A largely glazed base supports the whole building and contains the various shops. The studio worked specifically on the signage to unify the store signs displayed on the facade. The post and beam system chosen as the structural principle allows great freedom in the composition of the facades, which are each treated separately and adapted to their direct environment.
On the side of the Avenue de Neuhof, the building presents a tall and modern facade, dressed in stained concrete and punctuated by panels of dark wood clapboards. The facade states its verticality and consolidates the «urban» character defined along the avenue.
In contrast, the western facade has a more human scale and multiplies the transitional spaces between interior and exterior: accessible rooftop terraces, balconies and patios. This more porous architecture gives a feeling of intimacy in line with the adjoining apartment buildings.
The experience of the building is enriched by the lush greenery added to the project. The non-accessible roofs are planted with local and hardy species. Numerous green spaces and planted hedges are arranged on the first floor, mirroring the large trees that line the avenue.
The construction voluntarily implements simple and natural materials such as stained concrete, natural wood cladding, aluminium trims and hoods, etc. These choices demonstrate that it is possible to design a building with contemporary lines using sustainable materials and local resources.

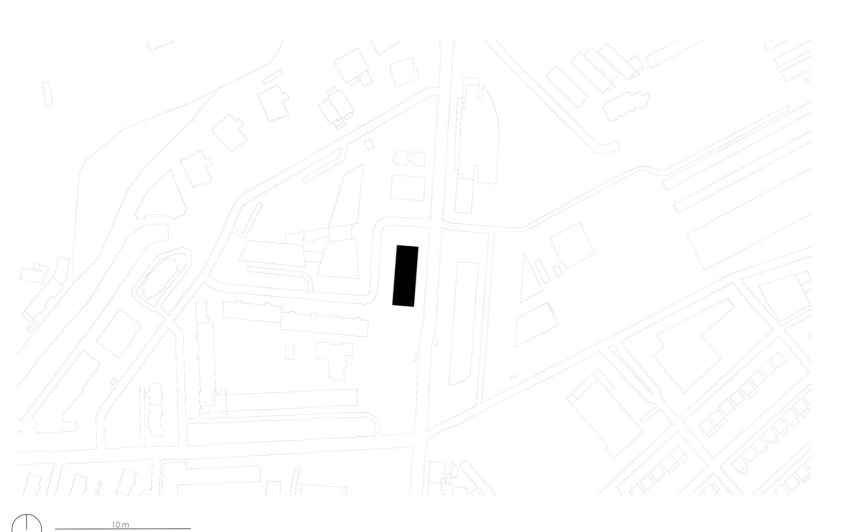

10 m

10 m

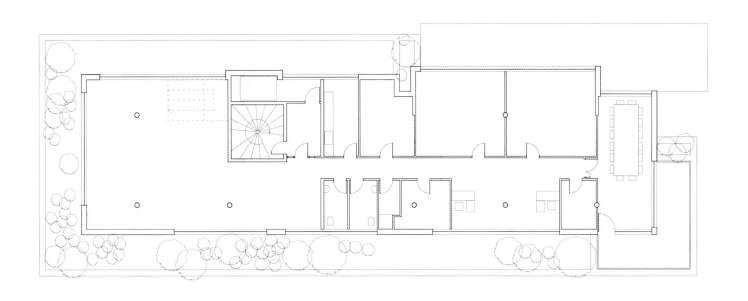

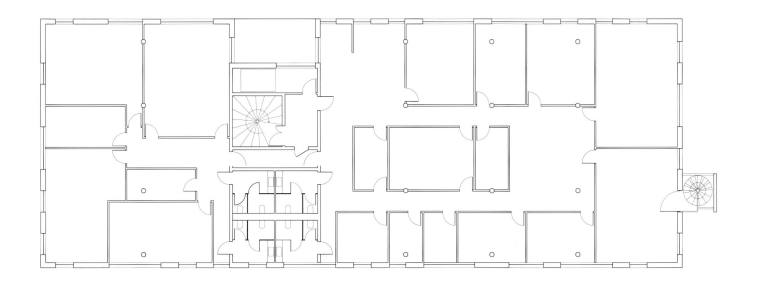

10 m

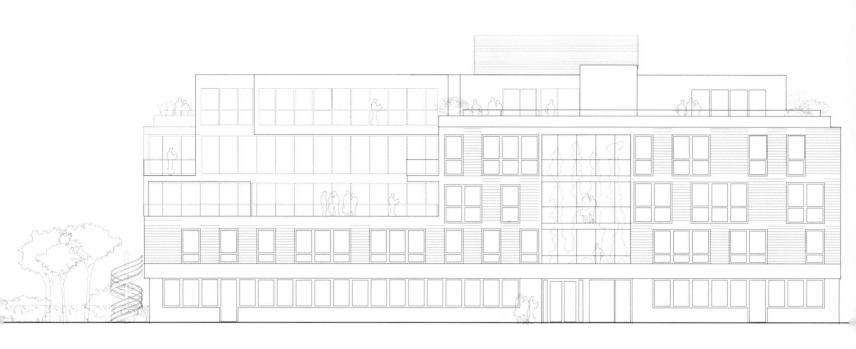

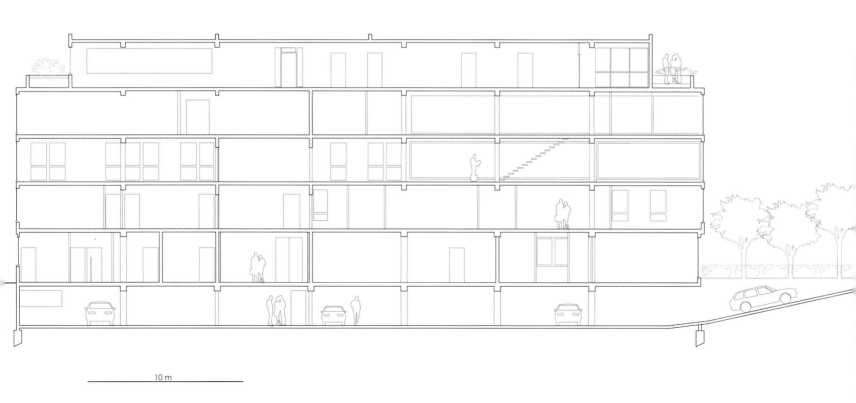

10 m

多媒体图书馆
Multimedia Library

2008年
埃尔斯坦 / 法国
Erstein / France

这里曾是一家毛纺厂，位于斯特拉斯堡南部，于2002年关闭。在此之前的150年时间里，其以独特的魅力塑造了埃尔斯坦小镇的面貌。2002年，城市开始实行棕地改造计划，以振兴该地区的工业遗产——将旧锅炉房改造成多媒体图书馆，标志着城市更新进程的开始。

办公楼的改造侧重历史背景并突出工业历史的教育意义。为此，原有的外墙和烟囱被保留下来（烟囱是引人注目的地标），让人能够情不自禁地联想到这里的历史。屋顶受损严重，建筑师通过特有的方式进行修复。建筑背面的倾斜结构被彻底拆除，但随后采用相似的结构加以替代。

通往建筑不同区域的路径更展示了其自带的工业化历史。例如，外立面上原来用于运送可燃材料的大栏杆被特意保留下来。建筑中央矗立着写有"Société Alsacienne de Constructions Mécaniques"（阿尔萨斯机械制造公司）字样的釉面砖锅炉。游客可以进入锅炉中心，一睹其过去的运行情况。保留下来的金属结构被漆成黑色，与柔和明亮的周围环境形成鲜明对比。

建筑内部原有布局未经任何改变，两层及三层高度的空间被保留下来。悬浮的透明吸音结构使光线可以透过屋顶照射进来，营造出温馨柔和的氛围。此外，夹层和活动室采用宽大的玻璃幕墙打造，便于欣赏运河以及室外的美丽景致——宽阔的庭院尽头似乎与远处的运河及大自然融为一体。

South of Strasbourg, the former woollen mill had shaped the different faces of the town of Erstein for 150 years before being definitively closed in 2002. Since then, the municipality has undertaken the conversion of the brownfield in order to revitalize the industrial heritage in the city centre. The transformation of the old boiler room into a media library marks the first step in this urban regeneration process.

The office's project focuses on the crystallization of the historical setting and on the educational restitution of its industrial past. The existing facades are preserved, as well as the chimney, a striking landmark recalling the history of the site. The heavily damaged roof is rebuilt identically. Only the lean-tos on the rear facade were demolished because of their poor state of conservation and replaced by similar volumes.

The path through the building multiplies the allusions to the industrial past of the site. On the outside, the large rail that used to carry the combustible materials is deliberately kept against the facade. In the centre of the building, the elegant glazed brick boiler, stamped « Société Alsacienne de Constructions Mécaniques », is made accessible, hosting small exhibitions. Visitors can thus enter the heart of the boiler room and get a glimpse of its past operation. The preserved metal structural elements are painted in black, contrasting with the soft, light atmosphere of the new layout.

From the inside, the vast volume of the building remains unaltered thanks to double and triple height spaces. Suspended acoustic sails contribute to creating a warm atmosphere and their diaphanous texture lets the light from the roof sheds filter through. With a generously glazed facade, the mezzanine and the activity room offer views of the canal and the outdoor spaces: a vast forecourt gradually descending towards nature and the Muhlkanal, which was diverted at the time to supply the spinning mill.

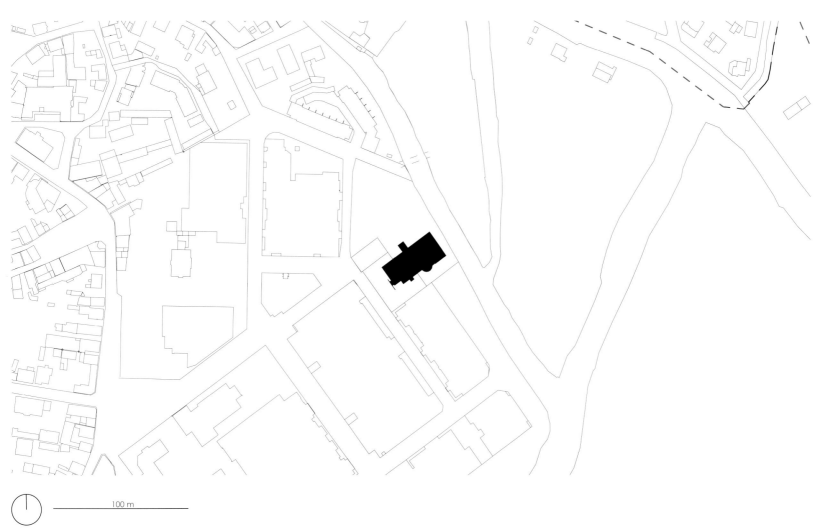

41

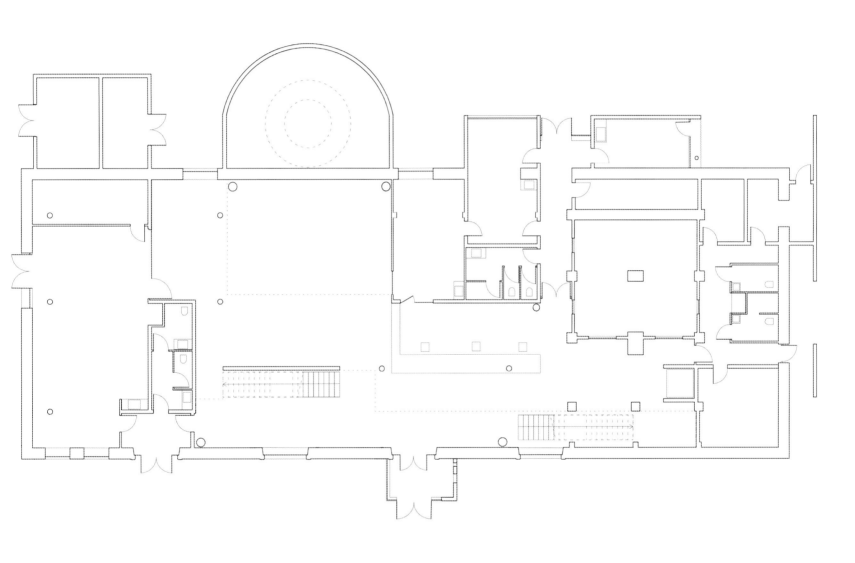

10 m

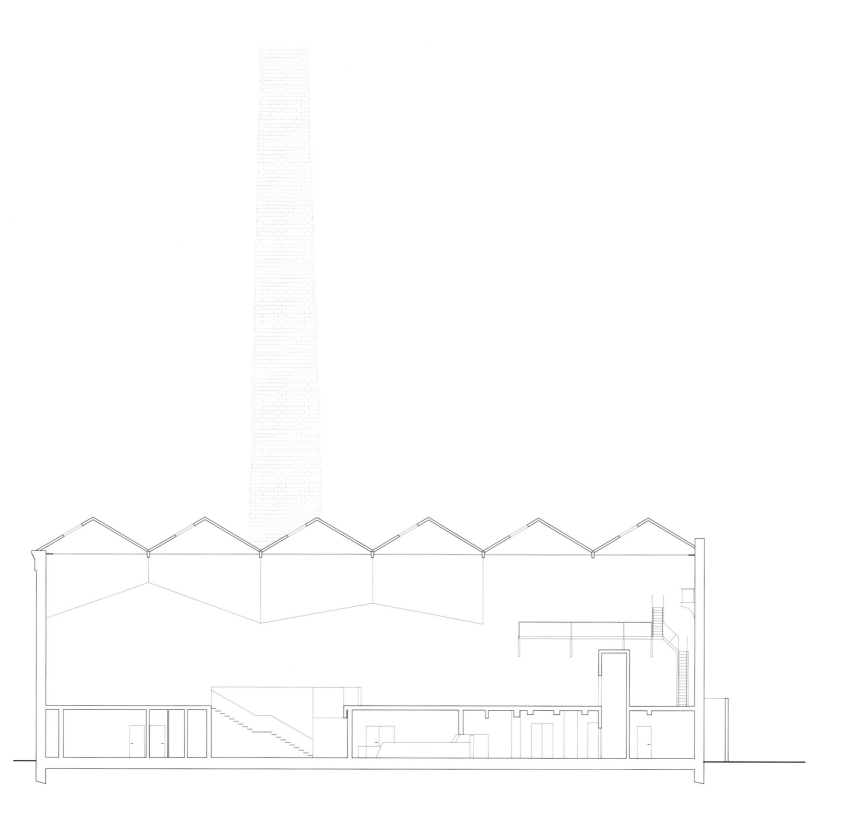

10 m

10 m

考古与科学博物馆
Archeology and Science Museum

拉巴特 / 摩洛哥
Rabat / Maroc

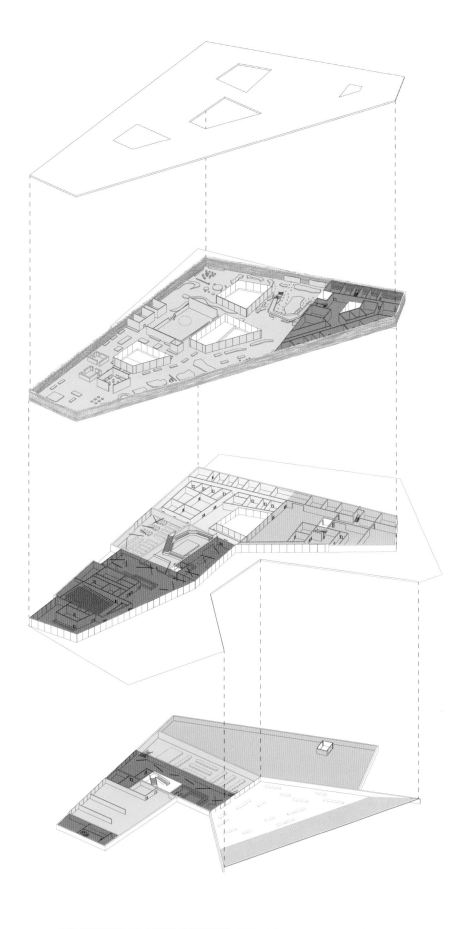

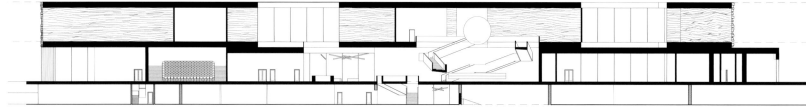

51

2010年

米雷库尔 / 法国

Mirecourt / France

农林业学校
Agricultural and Forestry School

米雷库尔农林学校位于城市南部人迹稀少的林区,其扩建项目选址在学校北侧,要求与现有建筑保持对齐。

设计不能打乱原有的地形,同时也不太能影响周围住宅楼的和谐氛围。因此,建筑师将新结构体视作原有楼群的延展——高中部分的新旧建筑通过天桥连接,但在视觉上保持独立。可以说,扩建部分在保留学校原有的风格同时,也为其创造了新的历史。

整体建筑采用回廊风格,围绕种植庭院展开。室内动线空间环绕花园打造,极具活力和生活气息。采用玻璃打造宽敞教室的立面,赋予空间极大的亮度和丰富的视野。需要格外提到的一点,花园是这个项目的重点——郁郁葱葱的植物环绕着中央池塘,营造出一片生机勃勃的景象。

整体建筑是对木材的致敬,也是对米雷库尔以及孚日地区在小提琴制作领域的宝贵技能的肯定。一层的楼板和顶棚采用产自该地

区的去皮树干作为主要支撑结构,这种材料超越其他技术,成为立面的主要特色。该学校是洛林地区第一所采用全木结构建造的高中。

这座建筑提供了一次真正的学习之旅:全木支撑结构随意倾斜,增添趣味性的同时更能引发对建设原则的质疑;花园由学生维护和管理;屋顶收集的雨水流向中央池塘。更深入地说,建筑的表达方式暗含了对森林的诠释,例如玻璃面板之间黄色和绿色的浮雕装饰。总之,所有的一切都是为了激发出学生的好奇心并唤起他们学习森林学的热情。

该建筑按照法国高环境品质评价体系打造,荣获2011年国家波巴特木材奖(公共和专业建筑类别)。宽大的屋顶悬垂形成遮阳板,保护建筑免受阳光直射。低辐射玻璃有助于控制进入建筑内部的光线。另外,室内动线空间充当教室之间的缓冲,并参与建筑整体的热调节。

The agricultural and forestry school of Mirecourt is situated in the south of the city, in a sparsely populated and wooded area. The extension is built on the north side of the school's plot, aligned with the existing buildings.

The new volume fits smoothly in the continuity of the existing, taking care not to disturb the topography and leaving the facades of the surrounding residential units unaffected. The old and new parts of the high school are connected by a footbridge but remain visually distinct. The extension creates a new layer to the school's history while keeping the original identity of the existing building unchanged.

In the style of a cloister, the building is organized around a planted patio. Circling the garden, the interior circulations are designed as fully-fledged living spaces. The generously glazed classrooms, open to the outside, providing great luminosity as well as views of a larger landscape. However, the garden remains the focus point of the project: lush and embellished with a central pond, it animates and illuminates the new building.

The architecture of the new high school is an ode to wood, a nod to Mirecourt's precious expertise in the field of violin making and to the Vosges region. The slabs of the first floor and the roof are supported by a row of debarked trunks from the region's forests. This wooden structure, highlighted and made visible, supersedes its technical condition to become the main feature of the facade. This is the first high school in Lorraine with a structure entirely made of wood.

The building offers a real educational journey: the random inclination of the wooden posts adds a playfulness to the build and raises questions about its constructive principle moreover the garden is maintained and managed by the students and the water collected in the roof runs off and feeds the central pond. Furthermore, the general architectural expression refers to the semantics of the forest: between the glass panes, tinted panels display a cameo of greens and yellows. Everything is shown to arouse the curiosity of the students, whose fields of study include forestry.

The building, built according to a High Environmental Quality approach, was awarded the Bobat Wood Awards prize (public and professional buildings category) in 2011. The large roof overhang forms a visor protecting the building from the sun. The low-emissivity glass also helps managing sunlight entering the building. Finally, the interior circulations act as a buffer space between the exterior and the classrooms, and take part in the thermal regulation of the building.

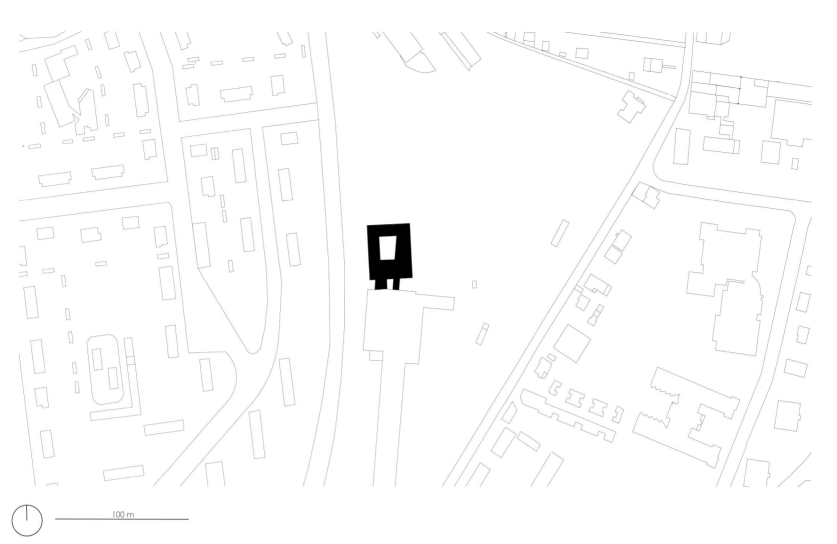

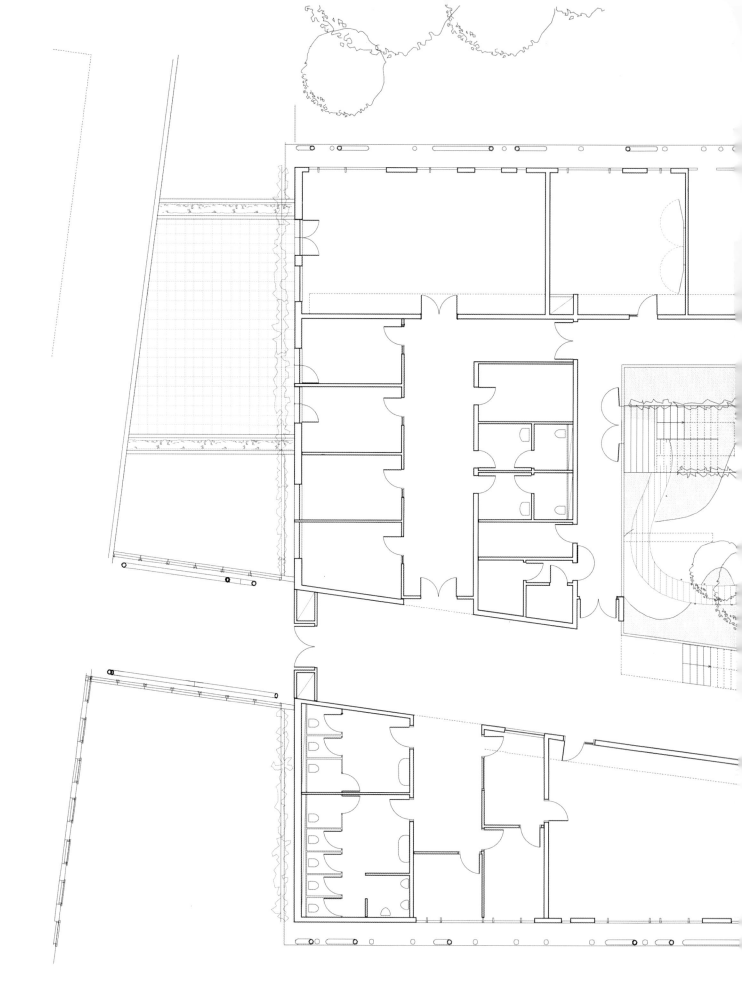

10 m

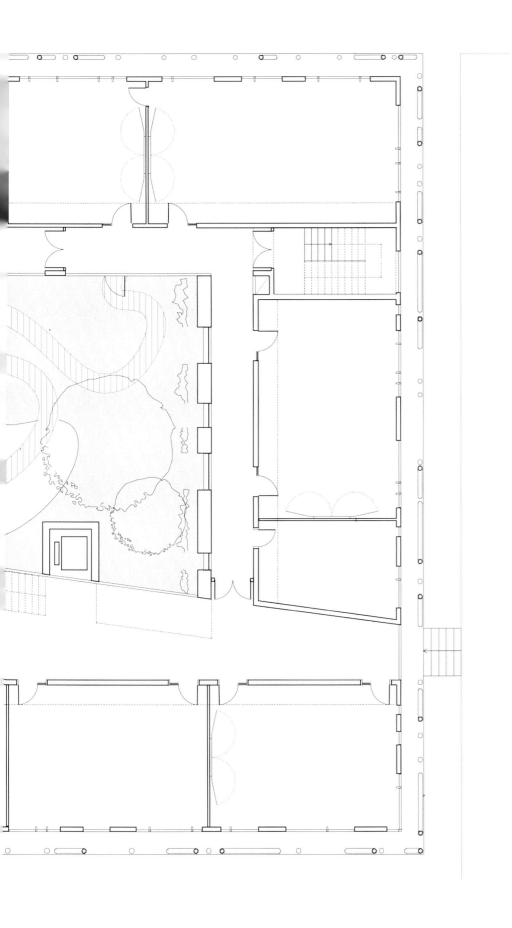

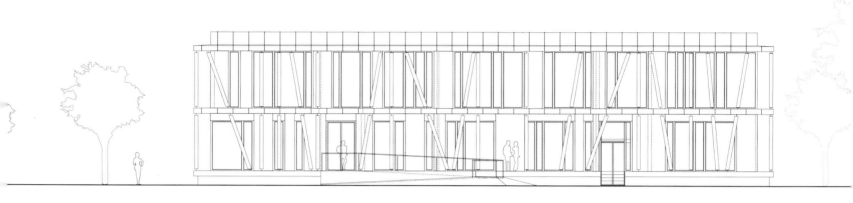

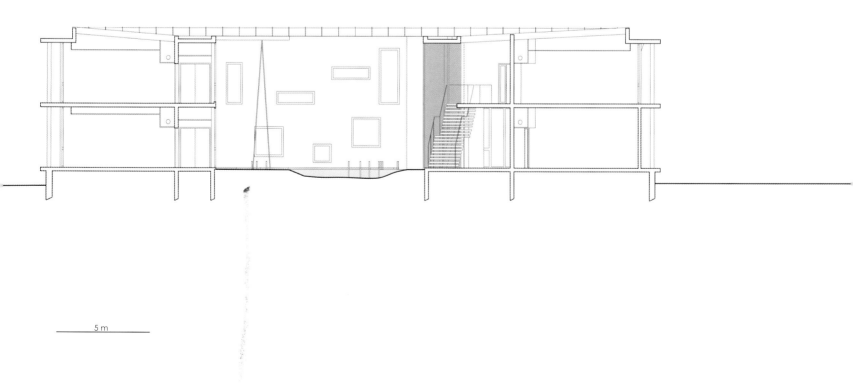

5 m

市政厅
Town Hall

2012年
诺尔杜斯 / 法国
Nordhouse / France

原有的市政厅太小且风格过时，因此业主要求打造全新的建筑。新市政厅选址诺尔杜斯小镇入口处、地标圣鲁丹大教堂前。

新市政厅体量紧凑且风格简约，既完美诠释出高效的施工流程，又与圣鲁丹大教堂及周围其他建筑交相呼应。建筑师选用的材料和色彩都与教堂相似，这使两座建筑能够和谐地实现对话，并打造出独特的小镇氛围。

整体建筑共3层，包括全高的接待大厅、等候区、办公区、会议室、茶室及服务场所等。建筑入口带有巨大的顶棚，这构成了建筑的主要特征。

小镇没有繁忙的路线穿过，从公路上看不太清楚小镇。建筑师充分利用其独特的地理位置来表明建筑的身份，进而借此来突出小镇的形象。整座建筑集合了传统市政厅的所有要素，诠释了功能性。建筑外立面采用高品质耐用材料打造——屋顶和南、北立面覆盖着铜板，东面和西面山墙覆盖着孚日砂岩板，呈棋盘图案。工整而垂直的开口增强了建筑本身的"管理"功能，用时钟和旗帜装饰的立面格外醒目。巨大的"NORDHOUSE"标志位于西立面底部，在进入小镇之前的交叉路口处就可以看到。

最后，接待大厅的地板上有小镇地图，游客穿行于诺尔杜斯的街道之间。

At the entrance of the town of Nordhouse, the new town hall project meets the need for a new building, the old one being outdated and too small. The new facility is located just in front of the Saint-Ludan Chapel, an important landmark in Nordhouse's landscape.

The new town hall echoes the architecture of the Saint-Ludan chapel, its simple and compact volume allows for an efficient construction process and harmonizes the building with neighbouring typologies. The range of materials and colours used are similar to those of the chapel. Facing each other, the two buildings dialogue harmoniously creating a structured town entrance with a strong identity.

The new building is built on three levels and includes a full-height reception hall, a waiting area, offices, a meeting room, a tea room and service premises. The entrance to the building is located under a large canopy, the only protruding feature of an otherwise compact volume.

A thorough reflection was carried out on the perception of the new equipment. The village of Nordhouse is not crossed by a busy route and is not very visible from the road. The institution therefore intends to take advantage of its new location to assert its presence and highlight the town's identity. The building reclaims all the codes of a traditional town hall to indicate its function. The building is clad with high-quality durable materials. Standing seam copper cladding covers the roof and the north and south facades. The east and west gable walls are clad in Vosges sandstone slabs laid in a checkerboard pattern. The rhythm and verticality of the openings reinforce the "administrative" aspect of the facility while a clock and flags adorn the main facade. A large «Nordhouse» sign is displayed on the base of the western facade, visible from the roundabout before entering the town.

Finally, the floor of the reception hall is covered by the map of the town, guiding the visitors through the streets of Nordhouse.

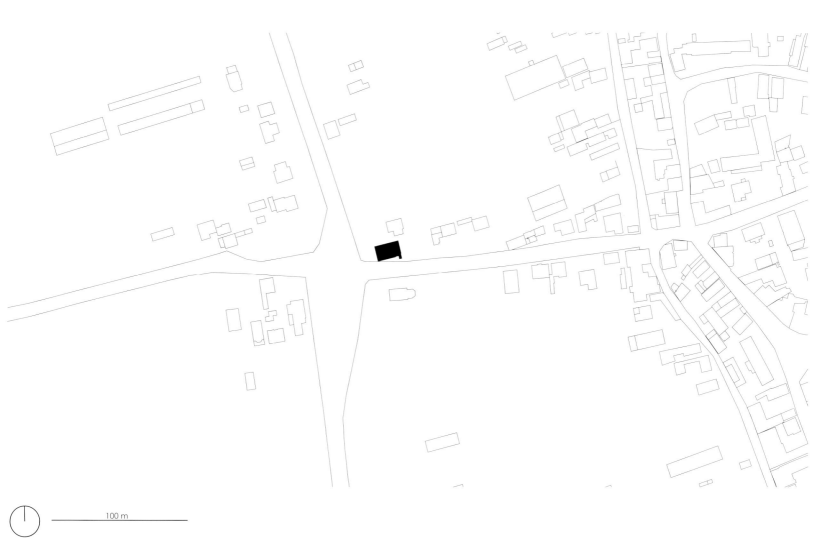

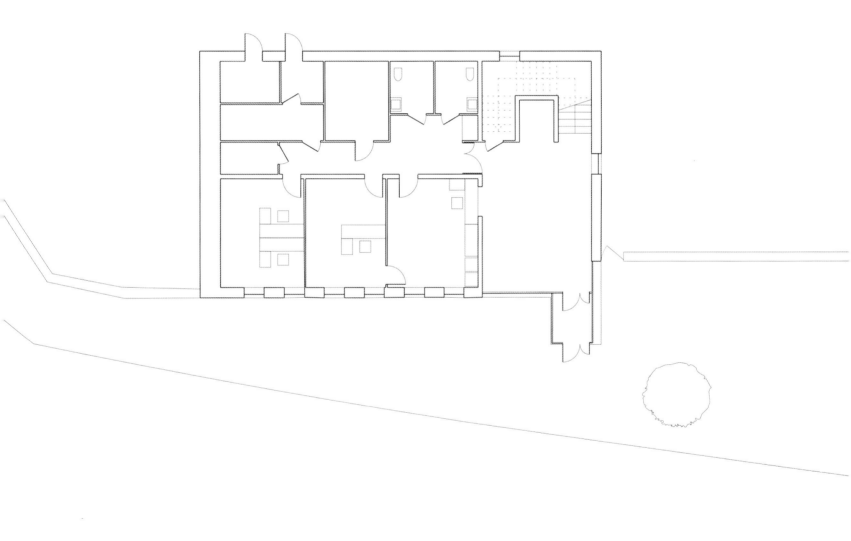

5 m

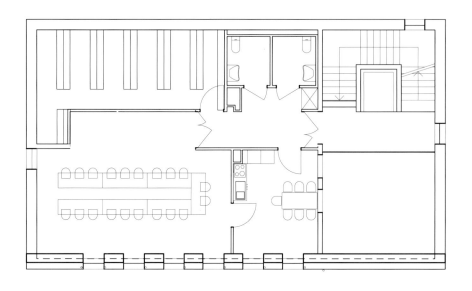

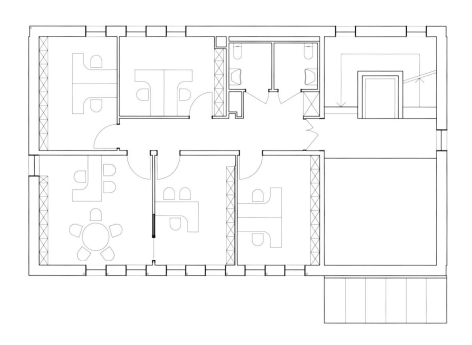

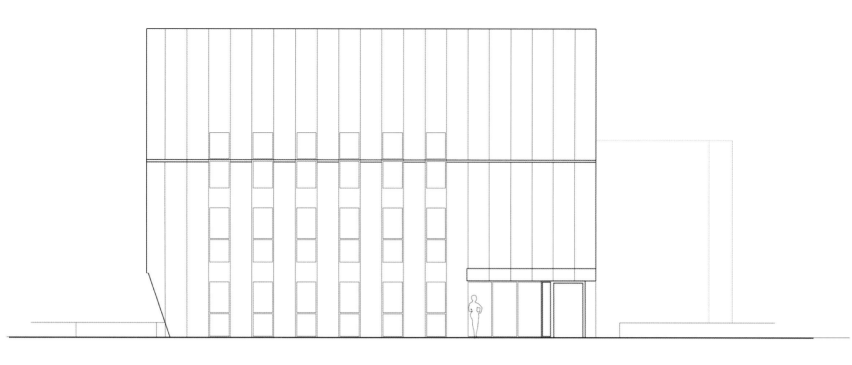

5 m

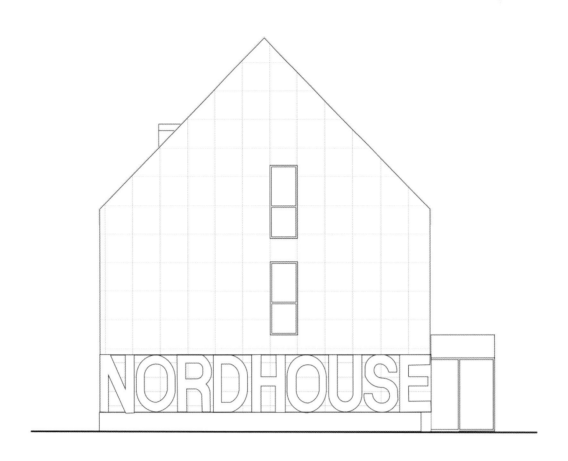

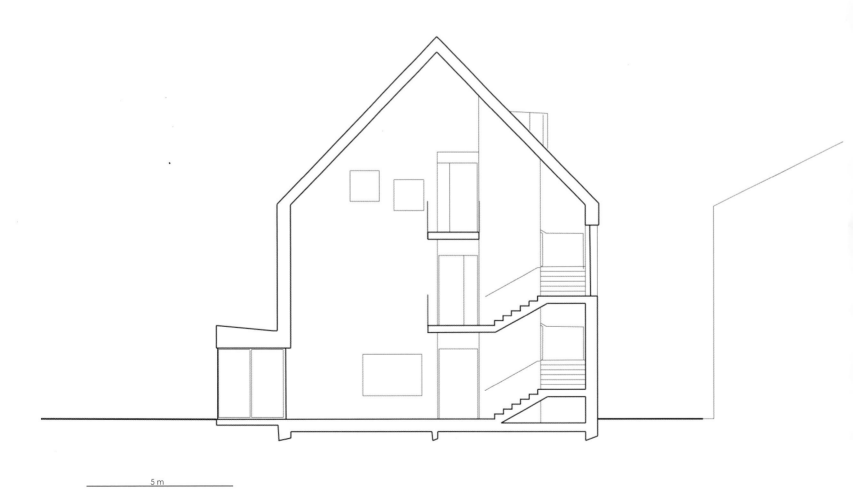

5 m

住宅
Housing

2014年
莱帕维永苏布瓦 / 法国
Pavillons-sous-Bois / France

该项目选址在莱帕维永苏布瓦"伊洛特运河城市开发区"内，独特的地理位置帮助创造了全新的城市发展序列。

建筑北立面传递出坚定自信的特色，巩固了蓬皮杜大道的新城形象。南侧整个街区朝向乌尔克运河（运河与阿姆斯特丹码头的人行道和自行车道相邻）。整个项目由75个住宅单元构成，独特的朝向使每一个单元光线充裕。7栋住宅楼围绕一个大型中心花园展开，建筑师尽可能将运河的美丽景致引入屋内。建筑高度随着地势坡度不断降低，直到最后打造出了一个适合休闲的健步长廊。

建筑群中心是一个美丽的阶梯式公共花园，一路引领着居民走到运河边，打造了一个令人愉悦的绿色户外空间。开阔的地面上种植着高大的树木，郁郁葱葱，别具特色。

建筑的不同部分通过外立面材料加以区分。底层采用黑色赤陶砖打造，增强了建筑本身坚固的形象，同种材料垂直延伸，凸显建筑的体积。水平布置的天然木材包层用于加深"突出"与"凹陷"部分的体量，其余部分全部采用白色饰面，打造出十足的对比效果。这些住宅楼大多是整层式结构，带有南向阳台，可欣赏公共花园、绿色屋顶以及运河的美丽景致。种植分隔墙以及抬高底层的方式确保了居民的隐私，并为他们创建了安静的生活环境。

该项目被评为BBC低耗能建筑，其采用被动式结构且安装了天然气供热系统，绿色屋顶收集雨水用于灌溉公共花园，自行车车棚以及人行道的规划鼓励人们采用可持续方式出行。

In the centre of the "Ilot Canal" urban development in Pavillon-sous-Bois, the project's plot contributes to the creation of a new urban sequence.
To the north, the assertive and vertical facades consolidate the new urban frontage of the Avenue Pompidou. Whereas to the south, the entire block opens onto the Canal de l'Ourcq, which is bordered by the pedestrian and bicycle path of the Quai d'Amsterdam. Thanks to this orientation, the 75 housing units of the project benefit from generous sunlight. The seven residential buildings are organized around a large central courtyard, offering as many views as possible towards the canal. Following the natural slope of the land, the height of the buildings gradually decreases until they reach a domestic scale more suitable for framing the pedestrian promenade.
The heart of the block is landscaped as a tiered communal garden guiding the inhabitants down towards the water and creating an enjoyable green outdoor space for them. Keeping open ground areas allows the creation of a real garden planted with tall trees.

Reading the different volumes of the built cluster is simplified by the choice of materials for the facades. Black terracotta bricks are laid on the ground floor adding a strong base layer effect to the project; the material extends vertically to display the structural volumes. A horizontal natural wood cladding is used to highlight the protruding and recessed volumes whereas white render unifies the non-clad volumes. The dwellings are mostly floor-through units and offer south facing balconies with views of the communal gardens, the green roofs and the canal. The privacy and serenity of the residents are guaranteed by creating planted borders around the project and raising the ground floor.
Thanks to all the solutions put in place, the project got the BBC (Low-Energy Building) label. The buildings are insulated from the exterior and equipped with a natural gas heating system. The rainwater is absorbed by the green roofs or collected to water the communal gardens. The project promotes the use of sustainable mobility by providing bicycle shelters and offering pedestrian pathways in continuity with the ones of the new "Ilot Canal" district.

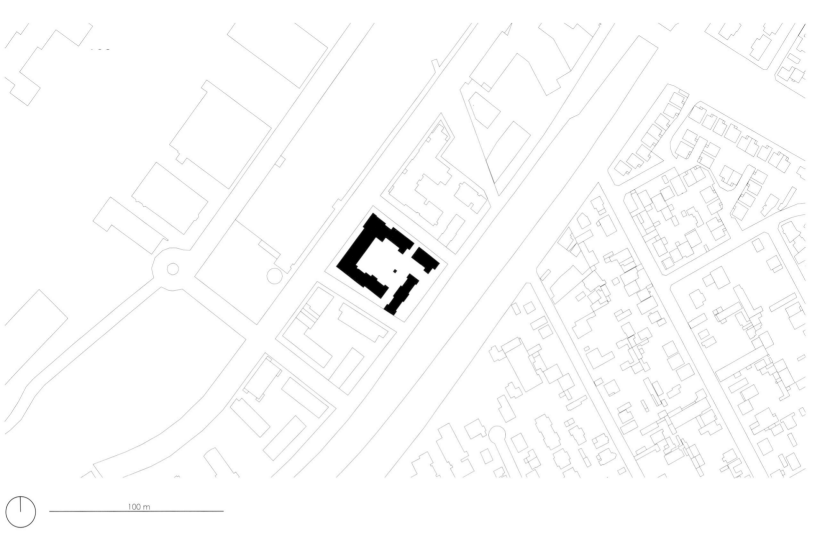

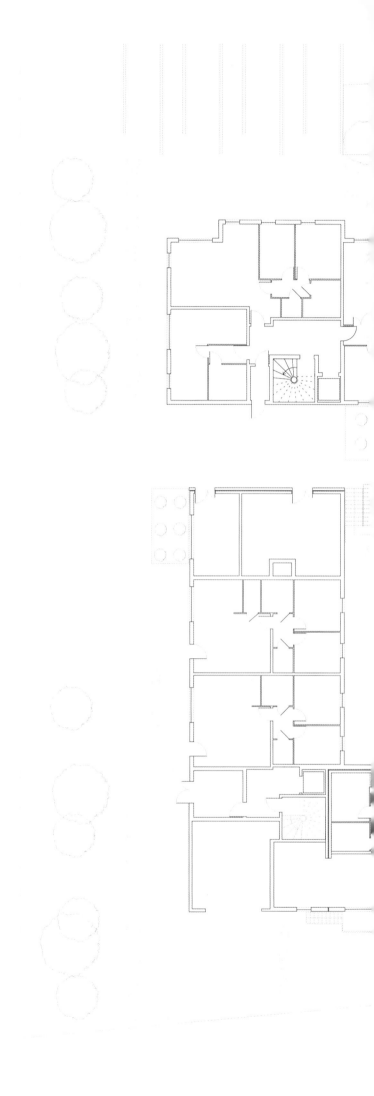

20 m

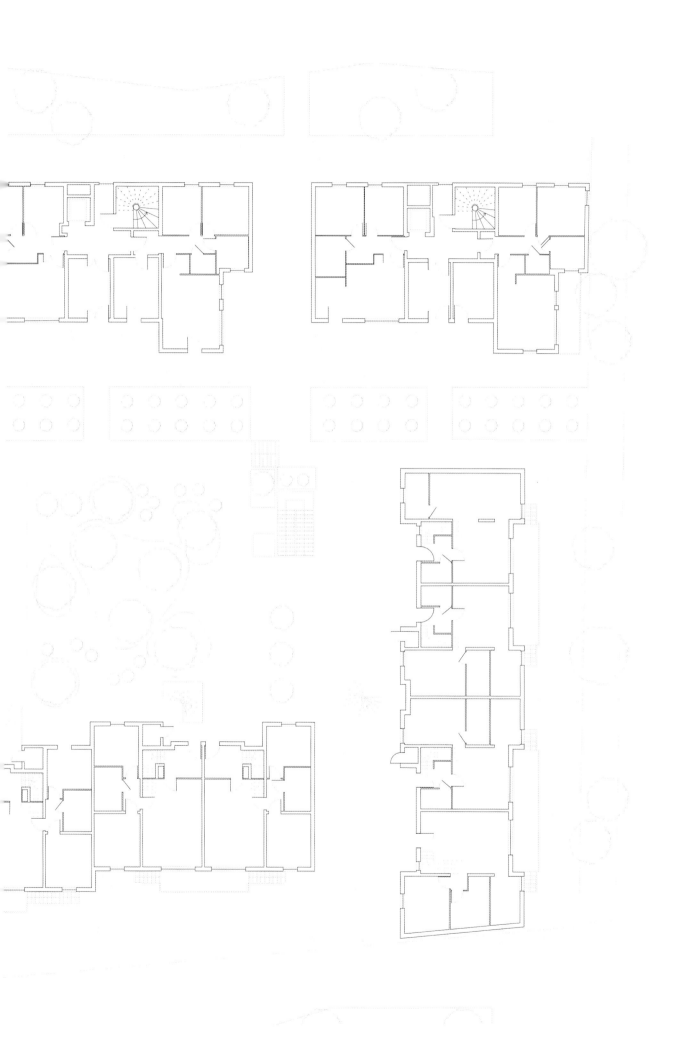

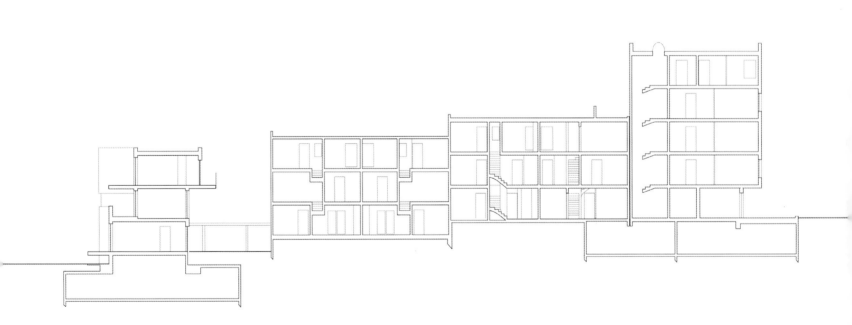

20 m

非洲领导大学
African Leadership University Campus

2017年
基加利 / 卢旺达
Kigali / Rwanda

这座建筑以可持续性为主要设计理念，着重考虑自然气候的控制。当地材质（如本地砖）有助于控制温度和湿度，独特的屋顶造型保护内部免受雨淋及阳光直射，同时也能确保空气的自由流通。

建筑施工的原则非常简单，即充分利用本地资源。其目的是认真打造未来的大学校园，创造可持续建筑。

This building is designed with sustainability in mind, and the focus is on natural climate control. Locally sourced materials such as brick help to control temperature and humidity. The roof protects users from the rain and sun, while allowing air to circulate freely.

The construction principle is very simple and highlights the assets of the area by making use of local resources. The aim is to conscientiously design the future university campus, laying great emphasis on the sustainability of the buildings.

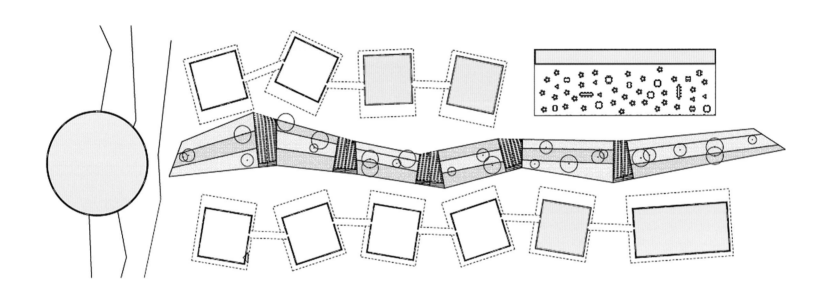

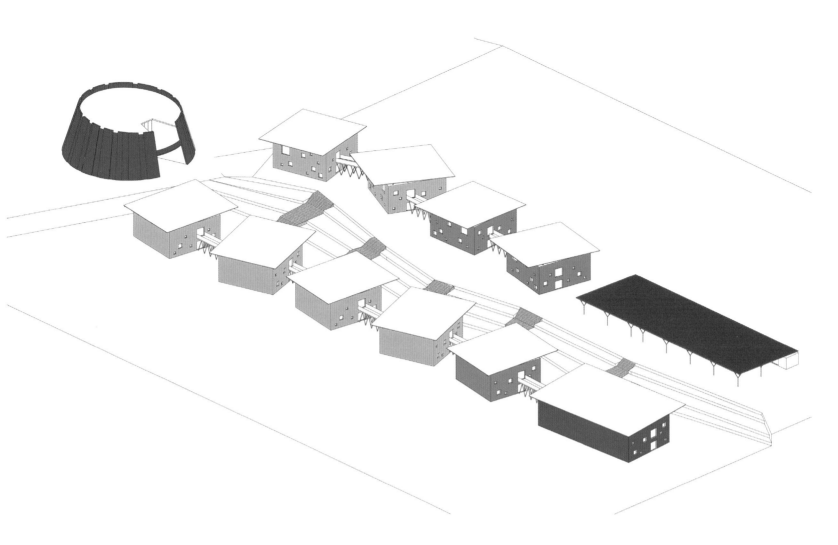

法国国立行政学院
National School of Administration

2015年
斯特拉斯堡 / 法国
Strasbourg / France

法国国立行政学院新校舍位于斯特拉斯堡市中心,其选址被列为历史古迹保护遗址。整个项目的设计和建设是在区域历史古迹保护员的指导下完成的。该地块是不同城市的交会点,情况相对复杂。为此建筑团队设计了感性的建筑群,既展示了对历史遗址的尊重,又体现了对与学科关联的教育法则的探究。

该项目的第一步是改造原有建筑的一层——已有的横梁和石墙被保留下来,与全新的家具和整洁的现代风格线条形成鲜明的对比。在建筑外立面上开凿出长长的带状窗户,让路人可以瞥见里面的场景。学校主入口依然保留在历史建筑内。

新建部分紧邻原有建筑,但看起来更加突出,且具有很高的可见性。两座建筑通过连廊相通,在这里可以从圣玛格丽特大街望向现代艺术博物馆。

新建建筑完美融入周围的环境之中,因此并不显得突兀。4个立面全部采用木结构打造,并采用玻璃板和电木涂层木板饰面,统一而规整。同时每个立面又各具特色,板材的布局随着周围环境及朝向而不断变化。

东立面安装了双层反光玻璃幕墙,映射出原有建筑与天空的影子,预示着过去与未来的交会。

西立面采用不锈钢网结构打造,更具现代感,与对面雄伟的公寓楼交相呼应。这面金属网被视作两栋建筑的隐私屏障。

北立面上栽种着植被,展示出建筑的环保特性,同时也构成了圣玛格丽特大街上的地标。

南立面上覆盖着光伏板,充分利用自然光,是对屋顶结构的完美补充。

建筑的每一处细节都是建筑团队经过深思熟虑的设计成果,以打造一个高质量的可持续结构,同时提升学生的生态意识。该建筑是法国首批获得绿色建筑评价体系认证的建筑之一。

The new premises of the Ecole Nationale d'Administration (France's prestigious civil service college) are situated in the heart of Strasbourg, on the site of the former "Commanderie Saint-Jean" which is listed as a protected historic monument. The whole project was designed with the consultation of the Regional Conservator of Historic Monuments. The plot is part of a complex area where different urbanities mingle. S&AA worked on a sensitive architecture, showing its respect for the architectural heritage and its desire of pedagogy that it associates to each of its projects.

The first step in revamping the school consist in the remodelling of the ground floor of the "Commanderie Saint-Jean". The existing beams and stone walls have been conserved and they contrast with the new furniture and its clean and contemporary lines. A long ribbon window is installed in the facade, granting the passers-by a small glimpse of the media library nestled inside. The choice was made to keep the main entrance of the school in the historic building.

The extension is placed next to the existing building in a way that allows it to stand out and to remain perceptible from all sides. The two buildings are connected via a first-floor walkway, allowing to see the Museum of Modern Art from the Rue Sainte Marguerite and to create a path leading to it.

The building earns its legitimacy within the site by connecting with all the surrounding elements of the landscape. The four facades of the extension, all timber-framed, are treated uniformly thanks to an alternating layout of glass panels and Bakelite-coated wood panels. Each facade does have its own identity, the panells' layout changes depending on the surrounding architecture and the orientation of the facade. To the east side, a double skin of slightly reflective glass panels covers the building. The facade vanishes and fades, intermingled with the reflections of the "Commanderie" and the sky. The past of the site is projected onto the screen of its future.

On the opposite side of the building, the west facade is adorned with a stainless steel mesh that gives it a contemporary and monolithic appearance, as a response to the imposing apartment building facing it. This mesh acts as a privacy screen for both facilities.

The north side features a planted facade that creates a signal from the Rue Sainte Marguerite and displays the building's environmental involvement.

Finally, to the south, the same layout integrates photovoltaic panels taking advantage of a sunny exposure and completing the ones on the roof.

Within this multifaceted architecture, every gesture is thought out to showcase a virtuous and sustainable architecture as well as to raise awareness about ecological issues among the students of the school. The building is one of the first in France to receive the NF High Environmental Quality certification for tertiary buildings.

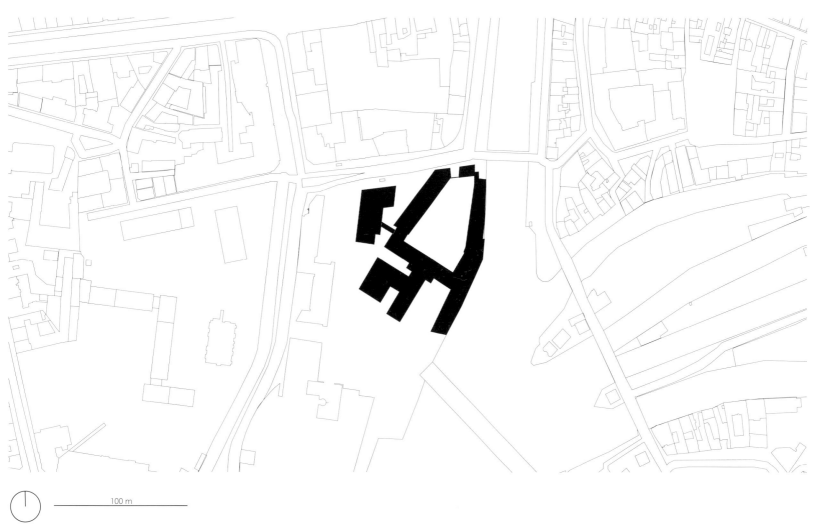

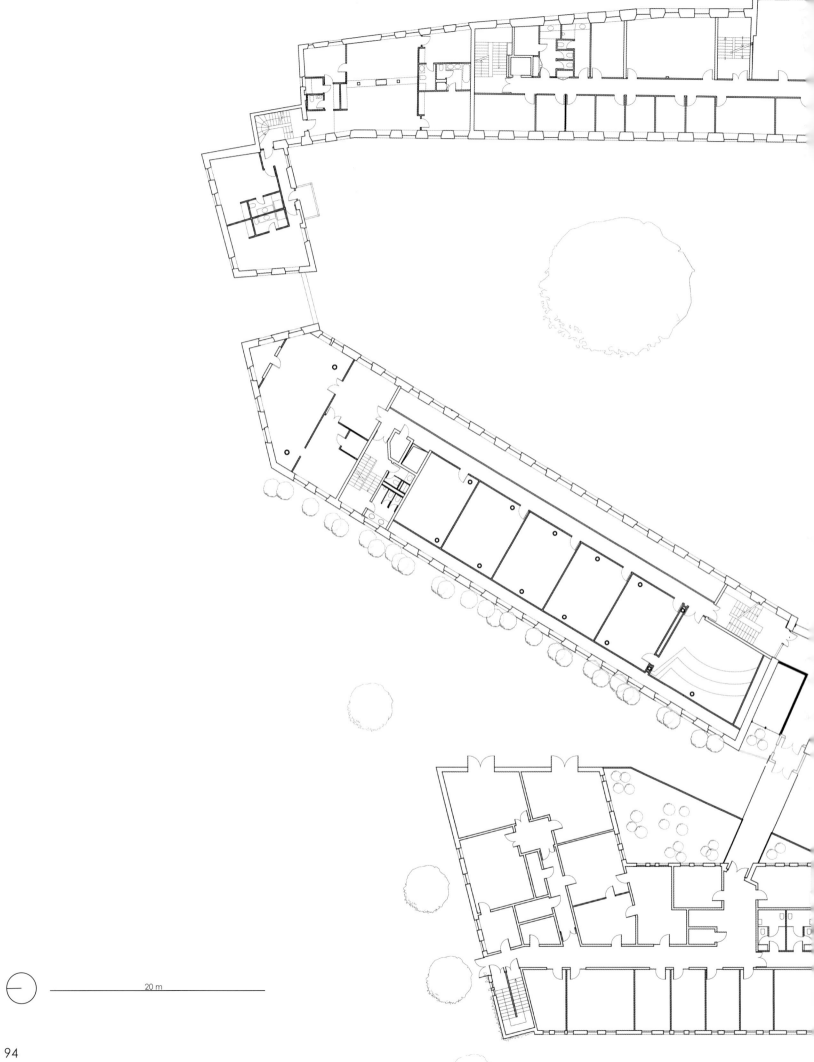

20 m

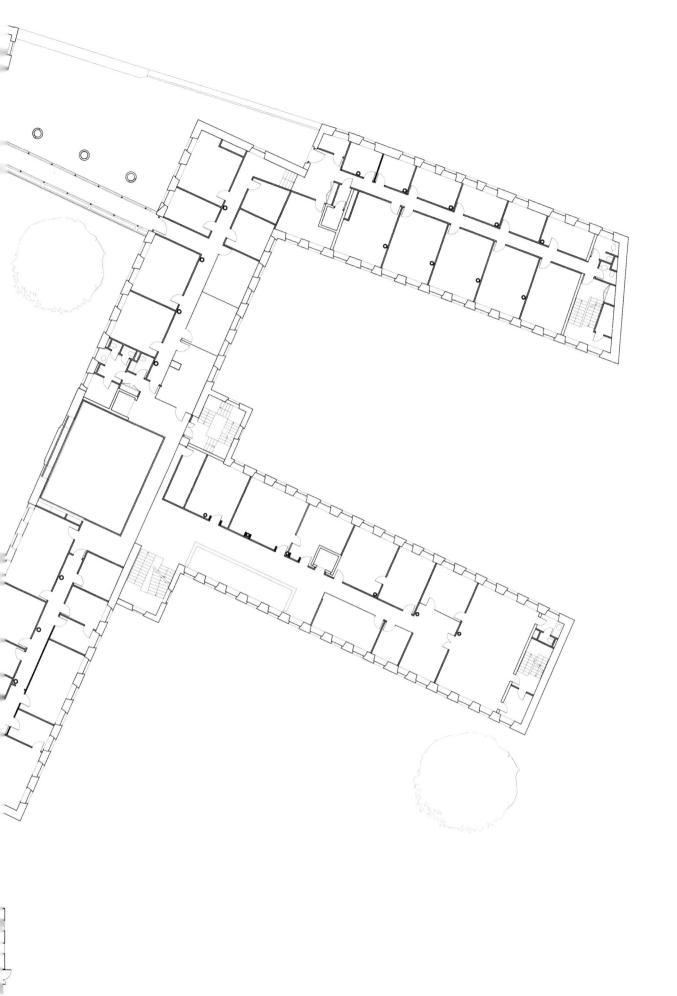

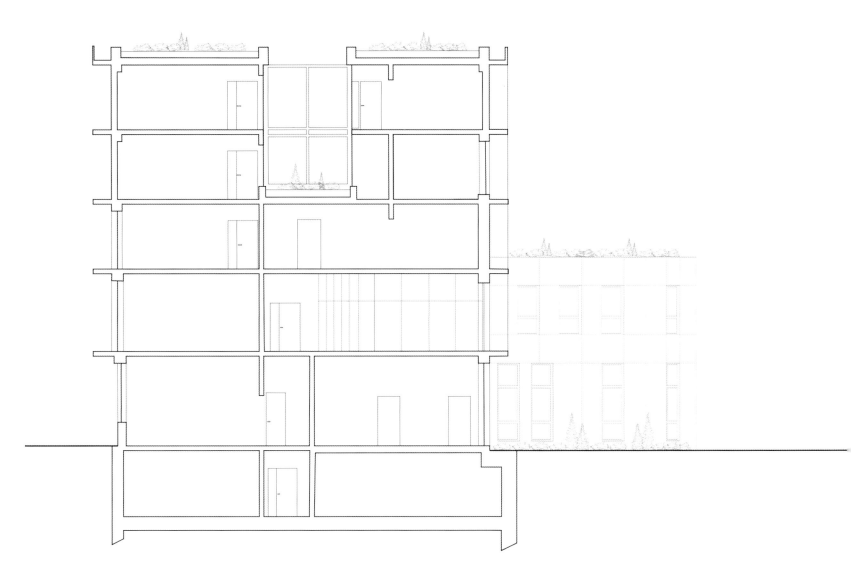

5 m

10 m

嘉士伯欧洲研发中心
Carlsberg European
Research & Development Centre

2014年
奥贝奈 / 法国
Obernai / France

这座壮观的建筑位于奥贝奈克伦堡啤酒厂（法国最大的啤酒厂）内，其象征意义十足的外观以及非典型的项目类型令人印象深刻——这是一个专为国际市场设计的新嘉士伯欧洲研发中心。

建筑场地达9000m²，包括一个用于全面测试新品的微型啤酒厂，一个2000m²的实验室，一个1500m²的存储空间，一个1800m²的办公区以及一个品尝区（由大小不一的小空间构成，用于每周测试开发的新品）。

整座建筑从501国道直接可见，位于新停车场、废水处理厂以及地下线缆区之间。为此，建筑的伸展和弯曲，顺应场地的独特条件。

垂直的木板条包裹着光滑的白色外墙，暗示着该品牌从斯堪的纳维亚起源。这层木板壳作为屏风，既可保护内部的隐私，又可避免外界的干扰，也可作为滤光器，保护内部不受阳光直射。动感的外观与嘉士伯（Carlsberg）标志的造型图案相互呼应，让人不禁联想到啤酒泡泡。

内部空间以创新为主要设计理念，现代、明亮而舒适，令人印象深刻。木板条后面主要采用玻璃结构打造，营造出开放的工作氛围。大堂和工作区采用壁画装饰，展示了集团的悠久历史以及位于维斯特伯的嘉士伯总部园区的美丽景观。

不得不说，这是一个非常独特的项目，建筑师玩起了教学式的游戏。沿路一侧立面上印着巨大的麦田图案，显示出周围的田野并描绘了酿酒过程中使用的麦芽。垂直的木板条代表着啤酒花栽培用到的支架。西立面的水镜诠释出啤酒酿造的另一种原料——水。这些直接唤起了人们记忆中与品牌和啤酒相关的词语。

随后，一个带有嘉士伯标志的倾斜的白色盒子从地面出现，清晰地标记了建筑的正门。而这个神秘的盒子是整座建筑的主要特色，被称作感官品酒室。

The spectacular "Zénith" building emerged from the ground in just a year, on the site occupied by Kronenbourg in Obernai, the largest brewery in France. The building impresses by its symbolic presence and atypical program: a research and development centre designing the new Carlsberg beers for the international market.
The 9,000 m² site includes a pilot microbrewery for full-scale testing, a 2,000 m² laboratory, 1,500 m² of storage space, 1,800 m² of offices and a blind tasting room where the different panels of ambassadors test the brand's new creations every week.
Directly visible from the departmental road 501, the building stretches and twists to find its place between the new parking lot, the water treatment plant, and the buried networks of a very constrained site.
Rows of vertical wooden slats are placed in front of the sleek white facades, alluding to the Scandinavian origins of the brand. This wooden shell acts as both screens and light filters, hiding the activities from the outdoor and thus giving a strong and secret aspect to the building as well as protecting it from too much sunlight. Its dynamic design echoes the curved motif of the Carlsberg logo, and brings

to mind the foamy aspect of the beer.
The interior spaces are impressively modern, comfortable and bright, and conform to the motto that governs the entire «Zenith»: innovation. Behind the wooden slats, the facades are largely glazed, and create an open workplace. The lobby and work areas are decorated with mural photographs depicting the Danish landscape of the Carlsberg campus in Vesterbro, the group's historic headquarters.
For this unique project, the office played the game of didactic architecture with straightforward evocations of the brand and the lexical field of beer. Along the road, a monumental barley field is printed on the building, reflecting the surrounding fields, and depicting the malt used in the brewing process. The vertical wooden slats refer to the hop growing supports. The last ingredient, water, is found on the west facade in a water mirror that leads the visitors to the entrance.
A sloping white box, bearing the Carlsberg Group's logo, emerges from the ground and clearly marks the building's main entrance. This mysterious box houses the main feature of the new centre: the sensory tasting room.

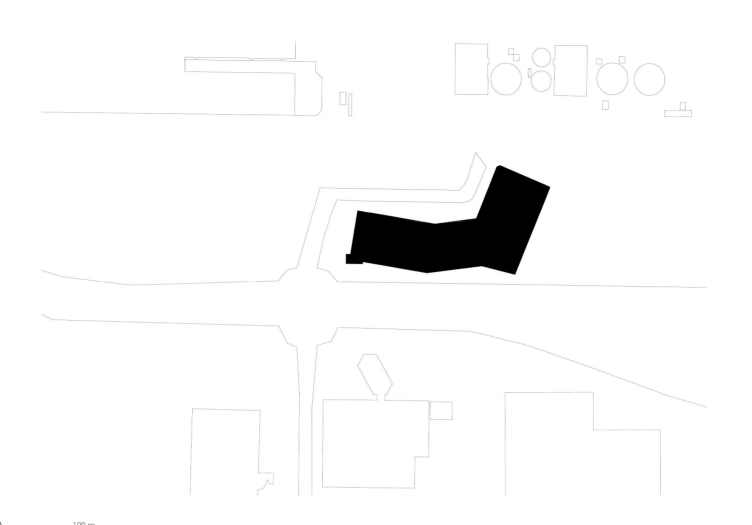

100 m

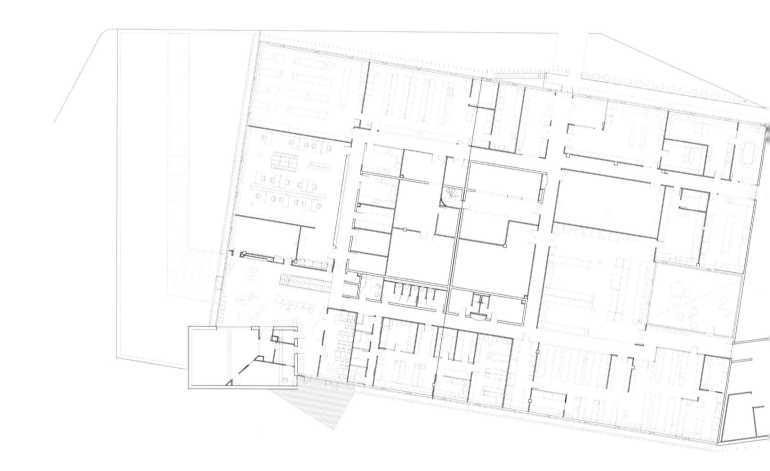

20 m

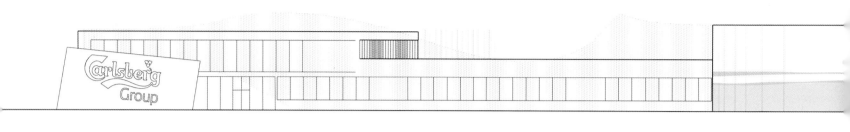

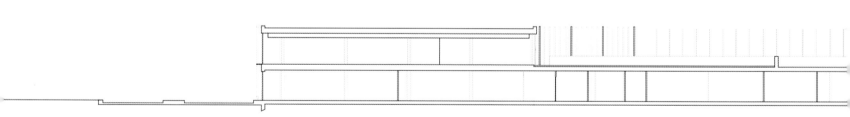

20 m

风湿病学研究所 / 综合诊所
Institute of Rheumatology / Polyclinic

2020年
达喀尔 / 塞内加尔
Dakar / Senegal

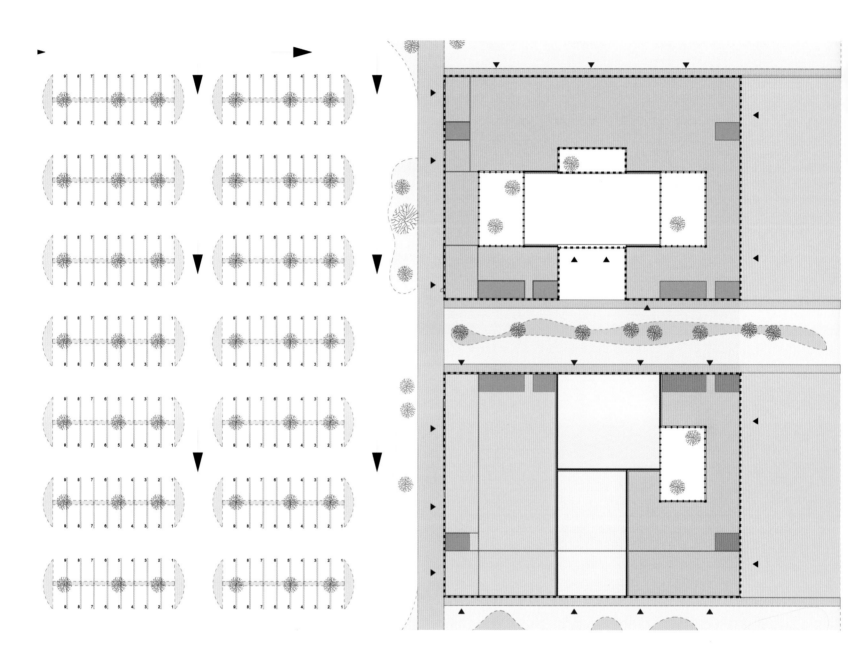

精神病院
Psychiatric Hospital

2014年

埃尔斯坦 / 法国

Erstein / France

埃尔斯坦精神病院专注于精神病学领域的研究和疾病治疗,其新建大楼蜿蜒曲折的外墙格外醒目,以其独特的姿态欢迎着前来的患者、医护人员和其他访客。

其设计理念源自"蜕变",如同毛毛虫变成蝴蝶的过程。客户要求设计一座规整且现代的建筑,既能够彰显医院自身的身份,又能够突出建筑特有的形象。角亭、大树以及沥青装饰带等共同构成了和谐的景观。

这座建筑由单独的小结构组成,犹如独立的展馆,通过中央接待大堂连接。蜿蜒的形态与原有地形交相呼应,并与周围景观融为一体。白色的外立面不断起伏延展,似乎在与四周的建筑嬉戏一般。

整座建筑符合BBC低耗能建筑标准,实现了医院推崇的环保理念,并为未来的其他设施的更新带来了便利。太阳能板用来加热生活用水,地下水热泵为建筑供暖。

医院的环境有助于患者的治疗康复。非线性的空间以及众多的开口便于营造充满曲线和光的氛围。另外,在设计中大量运用了环形图案,用于传递平和、丰富以及安全的感觉。这里是一个以病人为中心的地方,每个患者都可以找到属于自己的空间。这里也是一个充满活力的场所,让患者和医护人员如同是在聚会一般。整个项目从开始到实施一直秉承着"将接待、护理和生活融为一体"的宗旨。

医院绿化环境优良,满足了病人需要安然静养的需求。在小角落、花园以及停车场都栽种了植被。屋顶进行了绿化处理,花床可以阻挡路人贴近建筑外墙。

轻盈的建筑造型赋予内部空间足够的私密性与舒适度,让建筑与周围环境的关系和谐。

At the entrance of the Erstein Hospital Centre, specialized in psychiatry, the undulating facades of the new hospital building now welcome patients, nursing staff and visitors in a metamorphosed site.
The idea of metamorphosis, turning from caterpillar to butterfly, inspired the organic form of the project. The client wanted a balanced, contemporary building that would enhance the image of the site while maintaining its identity: a landscape of angular pavilions, tall trees and ribbons of asphalt.
The project, by its fragmented composition and the empty pockets that define it, reclaims the idea of separate pavilions while providing a central unifying reception area. Its flowing form weaves through the existing fabric and plays with the landscape, unfolding its white facade which undulates and flirts with the surrounding buildings.
Solar panels provide domestic hot water and a groundwater heat pump heats the entire building. The building thus complies with the BBC (Low-Energy Building) standard and sets the Erstein Hospital Centre on an environmentally friendly path as it will renew its other facilities in the future.
The entire project contributes to the care and therapy process. It avoids rectilinear spaces and multiplies the number of openings to create an atmosphere full of curves and light. The abundant use of circular patterns conveys a sense of calm, fullness and protection. It creates a place where the patient is at the centre, where security is assured without being dominant, and allows the patient to find his place. The space created is a dynamic space, a meeting place for patients and caregivers. It combines three main functions included from the start to the design process: reception, care, and living space.
The new hospital is nestled in a green setting to meet the patients' need for an escape. Numerous trees are planted throughout those nooks as well as in the gardens and parking lots. Vegetation also covers the roofs of the first floor and flower beds keep pedestrians away from the facades. The lightness of the building highlights the vegetal cover and amplifies the feeling of privacy and comfort within the building, as well as its integration into the site.

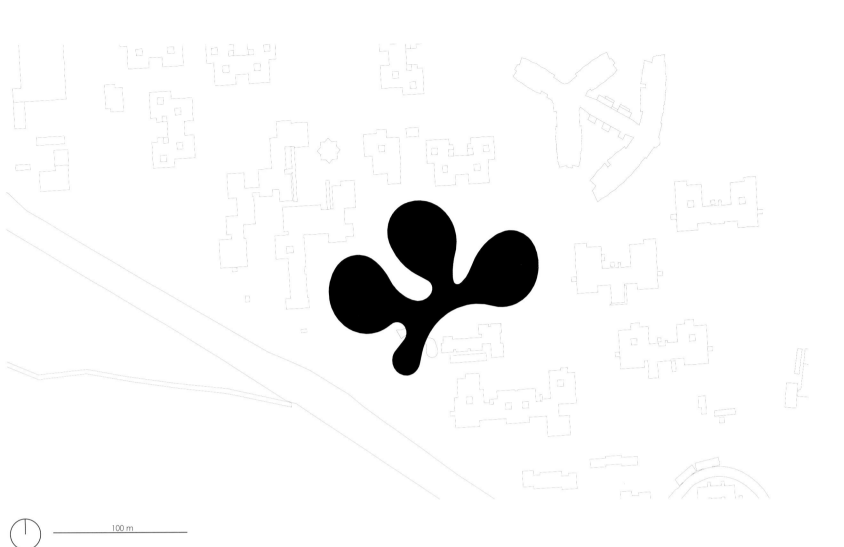

100 m

20 m

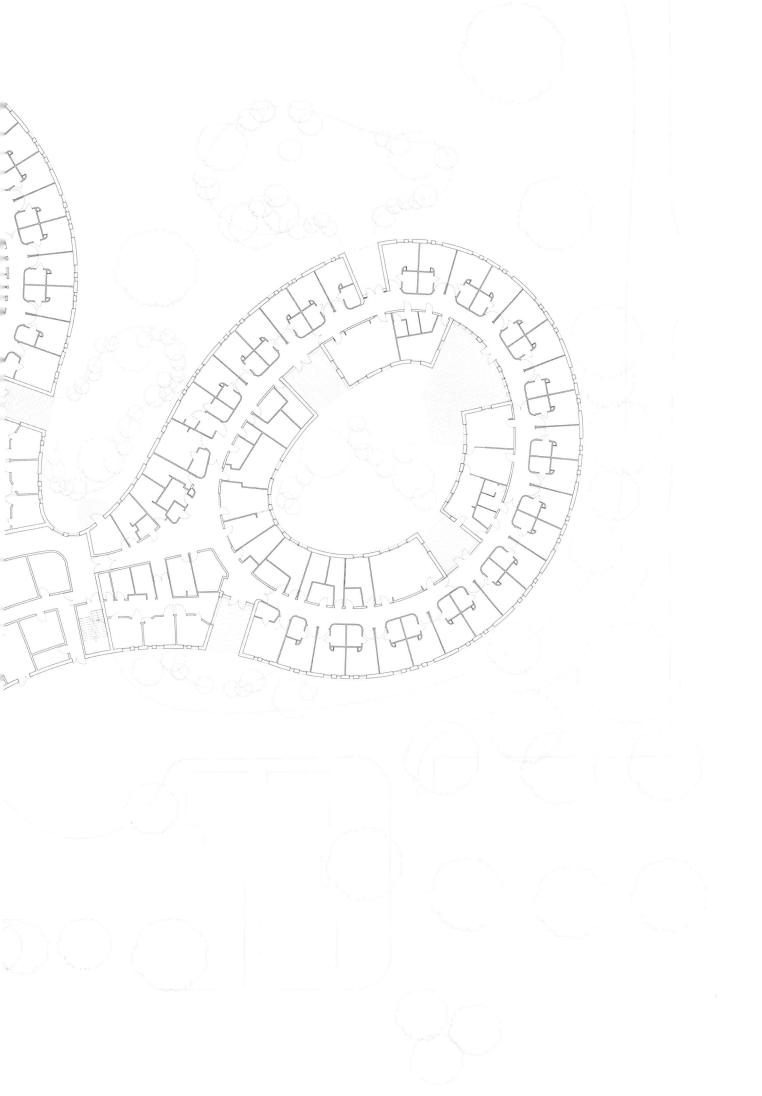

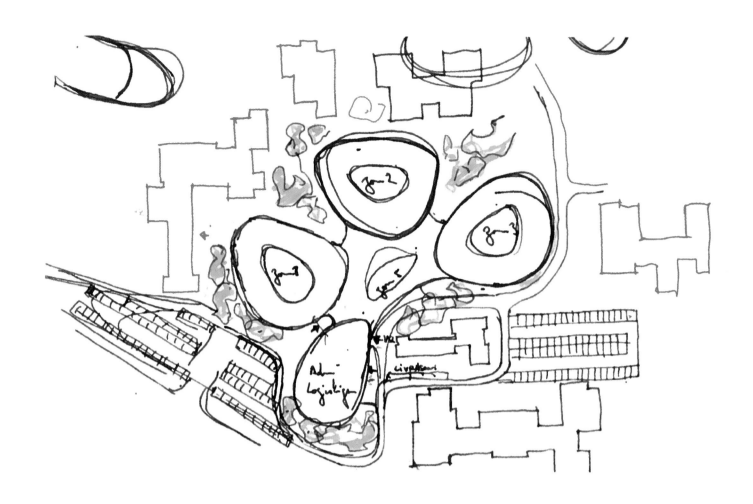

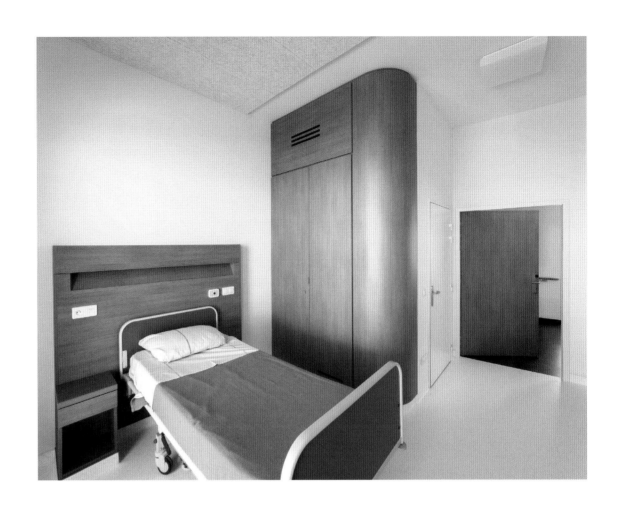

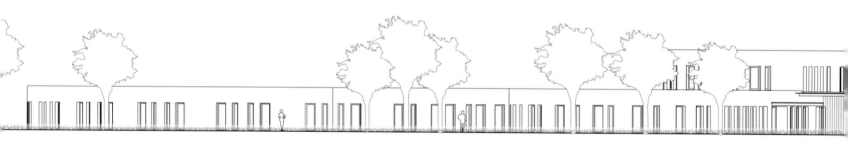

20 m

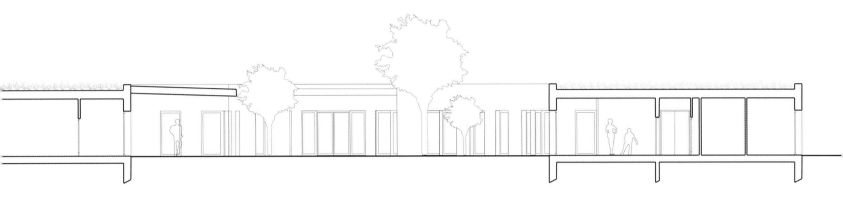

5 m

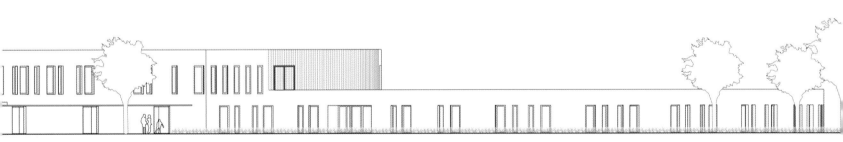

医院
Hospital

普法斯塔 / 法国

Pfastatt / France

2014年

普法斯塔 / 法国

Pfastatt / France

新医院大楼的玻璃幕墙宏伟壮观，映射出周围大街的美丽景致，格外引人注目。

该医院始建于1878年，为满足不断增加的需求，一直在不断地扩建和改造。如今这里包括一个多功能医疗中心，一个戒毒中心，一个老年病治疗中心，一个姑息治疗服务中心，一个老年人住宿疗养中心以及一个药房。

21世纪初期医院在进行扩建时暴露出了明显的问题：建筑结构缺乏层次感、统一性以及高效性。因此，必须对院区进行重组。基于这样的状况，设计团队除了要考虑内在品质之外，还要思考建筑场地的使用方式，使其与医院的现代身份更加匹配。

建筑师非常大胆地将新的医疗保健大楼规划在场地的北部，并使其与原有建筑保持平齐。这意味着为医院建立了一个新的形象与入口。其虽然与原有建筑分离，但是体量相似，巧妙地平衡了两者之间的关系。

新楼呈东西朝向，这使创建新的南北轴心空间成为可能，确保每个部门保持独立，又提升了整体操作流程。

新楼在室外创建了作为花园的连通空间，为病人和医护人员提供了舒适的休闲区域，同时也建立了室内地标。

南北立面构成了鲜明的对比。在北入口处，巨大的丝网印刷玻璃立面缓缓展开，其特有的光滑性和垂直的白色条纹增添了韵律感与活力度。这种抽象的物质性与原有建筑的外墙对比强烈。

南立面凸显俏皮而温馨的建筑语言，突出的结构打破了常规的开窗方式，并赋予楼下小花园更加人性化的气息。

随着扩建工程的不断开展，设计团队要在实施过程中考虑场地内所有建筑之间的关系。这座马奎萨德大街上的大楼换上了新的"外衣"，不仅提升了活力，而且在视觉上扩大了规模。

In Pfastatt, the imposing glass facade of the new hospital building draws attention and reflects the landscape of the Rue de la Mairie.

Inaugurated in 1878, the hospital has never stopped expanding and transforming itself in order to increase its capacity. Today, it includes a multi-purpose medical centre, an addiction treatment centre, a geriatric centre, a palliative care service, a residential facility for the elderly and a pharmacy.

However, when the project of a new extension appeared in the 2000s, the situation was clear: the disparate built fabric of the site lacked hierarchy, unity, and efficiency and the reorganization of the centre as a whole was necessary. The studio designed a project which, in addition to its intrinsic qualities, federates and rethinks the way the site is used, while matching the more contemporary identity of the hospital.

The new healthcare building is boldly placed on the northern part of the site, parallel to the historic building. This layout establishes a frontality that asserts the new northern entrance. Detached from the historic building, the extension appears as a separate volume similar in size to the existing building, balancing out their relationship on the site.

The east-west orientation of the new building makes it possible to create a north-south main axis. This previously non-existent axis structures the whole site, improving the flow of its operations and making each department independent. Outdoors, the presence of the new volume creates in-between spaces that are laid out as gardens. These living spaces enhance the comfort of patients and medical staff, and create landmarks from the inside.

The north and south facades contrast in an antithetical manner. By the northern entrance, a long monolithic facade clad in silk-screened glass unfolds. The fluid and vertical design of the silk-screened glass rhythms and energizes the facade with white stripes. This abstract materiality contrasts with the facades of the historic building.

The south face is designed with a playful and inviting language. Protruding volumes disrupt the regular pattern of the openings and bring a more human scale to the gardens established on the ground floor.

Along with the extension, the restructuring process tackles all the buildings on the site. The historical building is entirely refurbished and the building dating from the 70's, along the Rue des Maquisards, is wrapped in a new insulating skin to improve its energetic performances and increase its size.

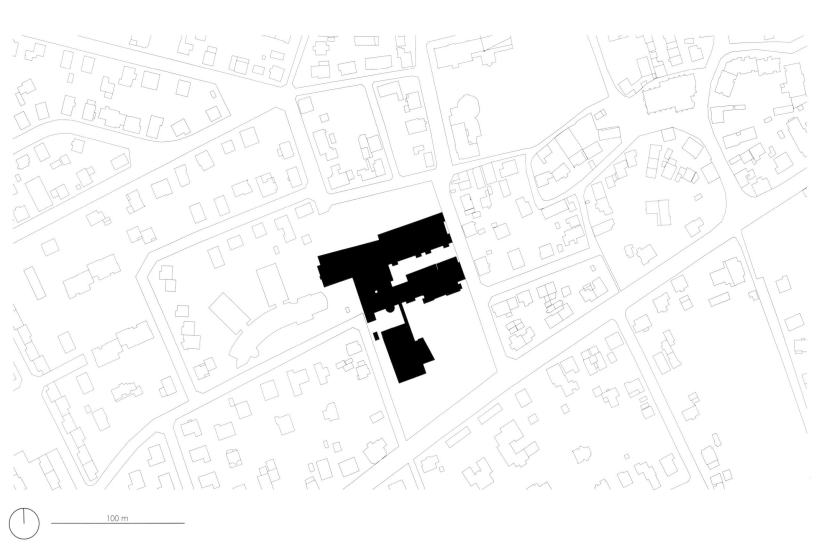

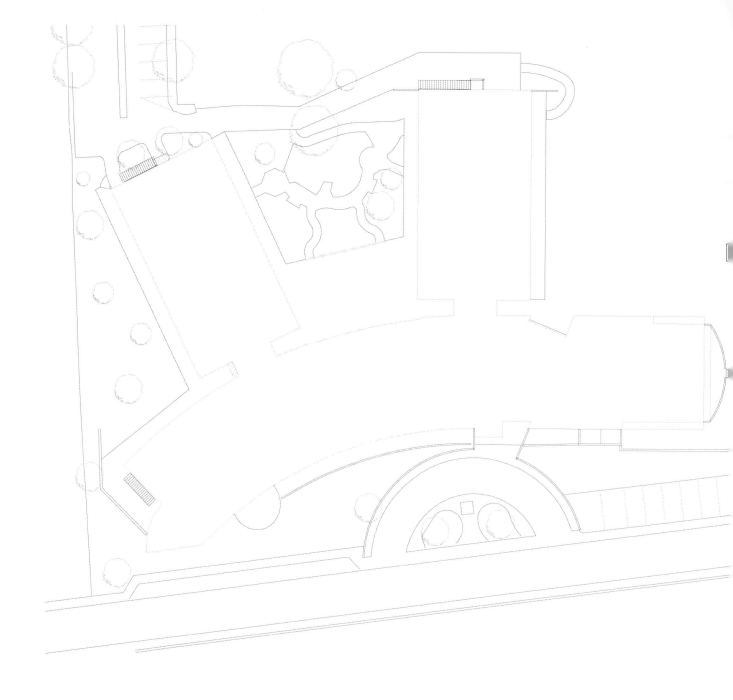

137

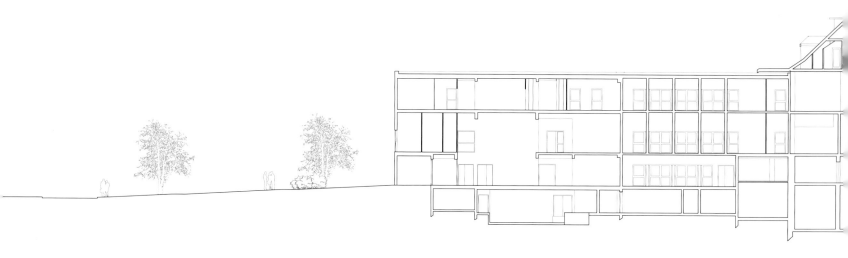

20 m

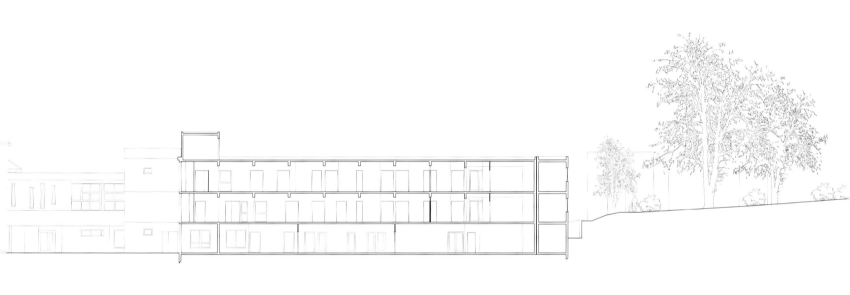

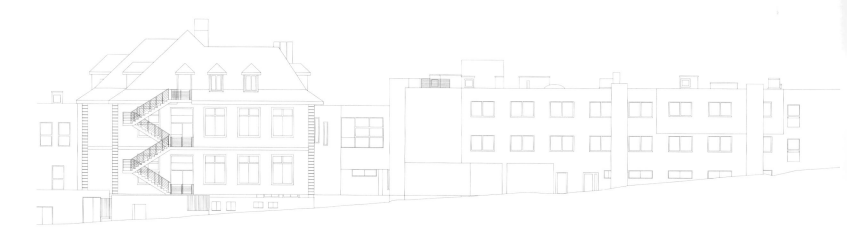

20 m

康复中心
Nursing Home

2014年

奥伯罗斯伯尔让 / 法国

Oberhausbergen / France

伯特利健康中心扩建工程包括一个养老院(退休之家)和两个供阿尔茨海默病患者生活的居住单元。"漫游"是阿尔茨海默病常见的症状,也是该项目的主导理念。为此,建筑团队专门打造了一条安全、宁静的无障碍道路,方便居住者的生活。

扩建楼为3层矩形结构,由两个天井贯穿,并以回廊形式呈现。每一层的走廊都围绕花园而建,房间朝向室外。在这里,所有的空间都是弯曲的,明亮而流畅,确保居住者带着愉悦的心情"漫游"。

两个花园之间的建筑内有会议室和餐厅,与楼上的露台连通。这里采用全玻璃材质打造,便于护理人员密切关注发生的一切。

扩建的养老院位于整个场地的尽头,旁边是一片田野,这里曾经是一个花园。为使建筑与周围自然景观相协调,设计团队专门打造了一个绿意盎然的生活环境。

白色的外墙采用开花植物及常绿攀爬植物覆盖,天井被设计成小花园,供居住者栽种植物或休闲。建筑外部的空地被打造成教育园地。

该项目设计还包含一项特殊的要求,即打造一个冥想室,用于冥想或礼拜,方便每一位有信仰的居住者使用。

冥想室被赋予简单、朴素、精致的风格,在这里精神成为主导。内部采用轻质木材饰面,墙体底部用玻璃打造,使整个空间仿佛悬浮在空中。从天花板辐射出的光制造了一种神秘而温暖的气氛。

整个房间犹如一个"茧",鼓励使用者摒弃外界的干扰,集中注意力进行内省。半透明的门窗允许光线进入,家具采用同种木材制作,包括一张桌子、两个架子和一条长凳。

冥想室获得了当地"木建筑"奖。

The expansion of the Bethel Health Centre includes an EHPAD (Retirement home) as well as two secure living units for people with Alzheimer's disease. The idea of "wandering", which is a frequent symptom of Alzheimer's disease, is the leading concept for the design of the project. The new building contributes to the well-being of its residents by providing a safe and serene pathway without obstacles. The expansion takes the shape of a three-storey rectangular building pierced by two patios and operates in the manner of a cloister; on each floor, the corridors surround the garden, while the rooms open up onto the outside. Inside, everything is curved, well-lit and fluid so as to ensure that the residents' restless wandering becomes a pleasant and soothing stroll. In between the two gardens, the building contains the meeting and dining rooms and opens up onto the garden or large terraces on the upper floors. From this central, fully glazed area, the nursing staff can keep an eye on the residents as they move around.
The retirement home's extension is built at the end of the plot, alongside a field, on a piece of land that was previously occupied by gardens. The strong presence of vegetation encouraged the studio to design the project in harmony with the landscape in order to offer a very green living environment.
The exterior white and sober facades are covered with flowering and evergreen climbing plants. The patios are designed as two small gardens in which patients can garden or simply enjoy a pleasant and safe space. On the outside of the building, the leftover land is redesigned as educational gardens.
The programme of the project features a special request: the creation of a meditation room which would benefit to all the residents of the Health Centre. This space is devoted to meditation or worship, regardless of the residents' beliefs. The studio suggested a simple, plain and very refined volume, so as to allow the spirit to take complete possession of the room. Inside, the interior is entirely clad in light wood. The four walls that structure the volume, with glass at the base, seem to levitate. A mystical and warm atmosphere radiates from the overhead light source that marks out the ceiling.
The room is designed as a cocoon; views to the outside are deliberately forbidden in order to encourage concentration and introspection. The translucent entrance door allows only light to pass through and limits all transparencies.
A monastic furniture, made of the same type of wood, enriches the room's purpose: a table, two shelves and a bench encircling the space.
The meditation room won the regional wood construction award.

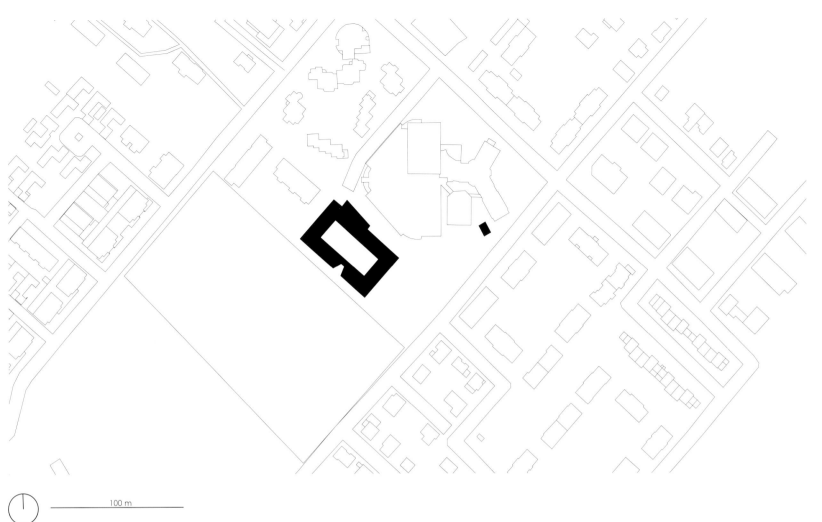

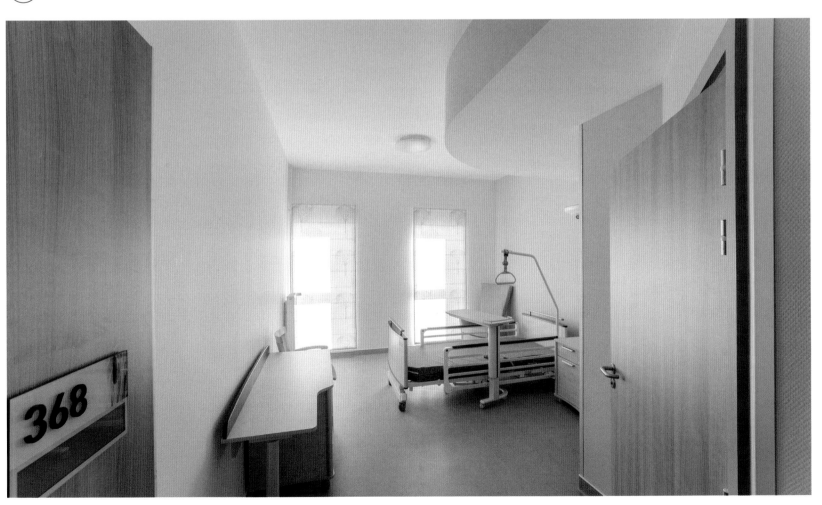

20 m

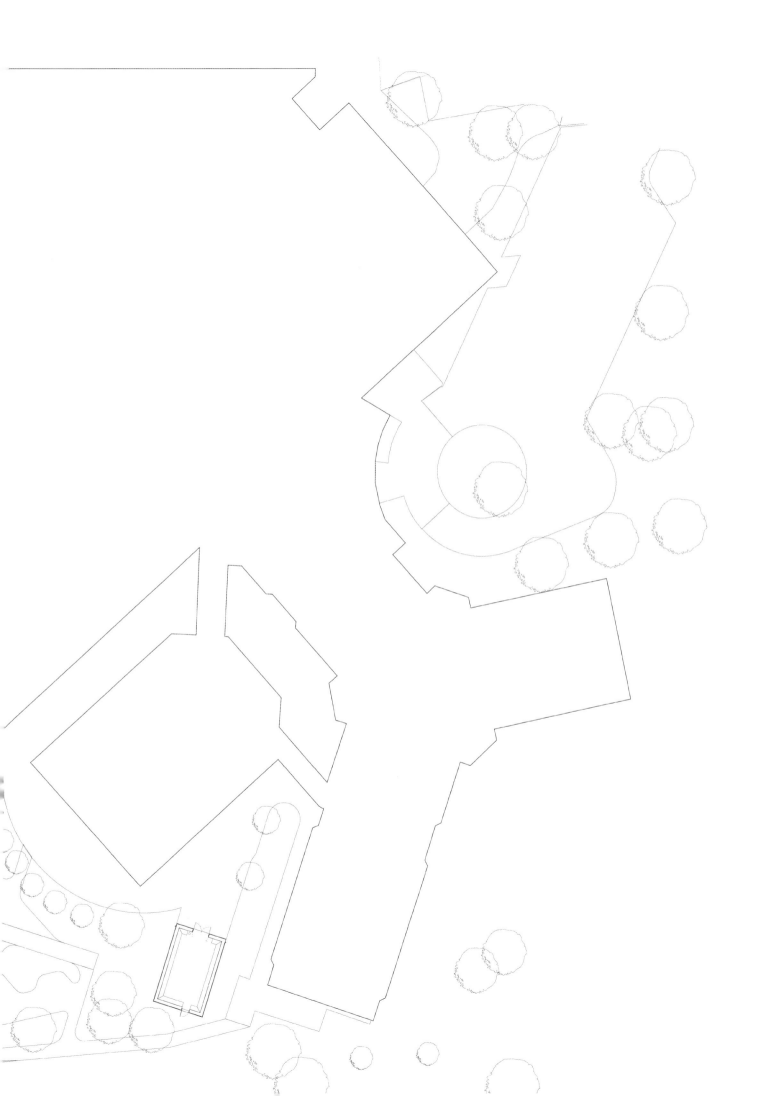

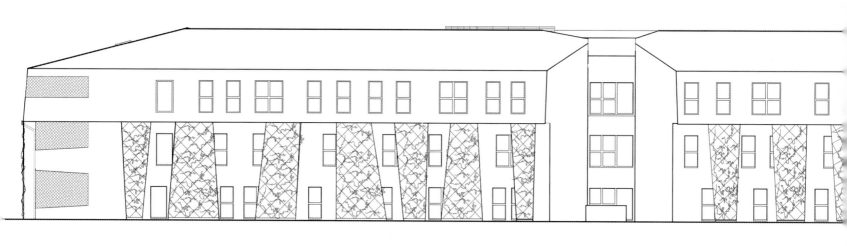

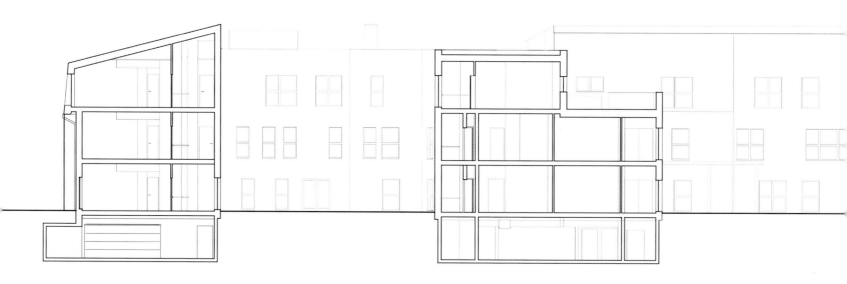

10 m

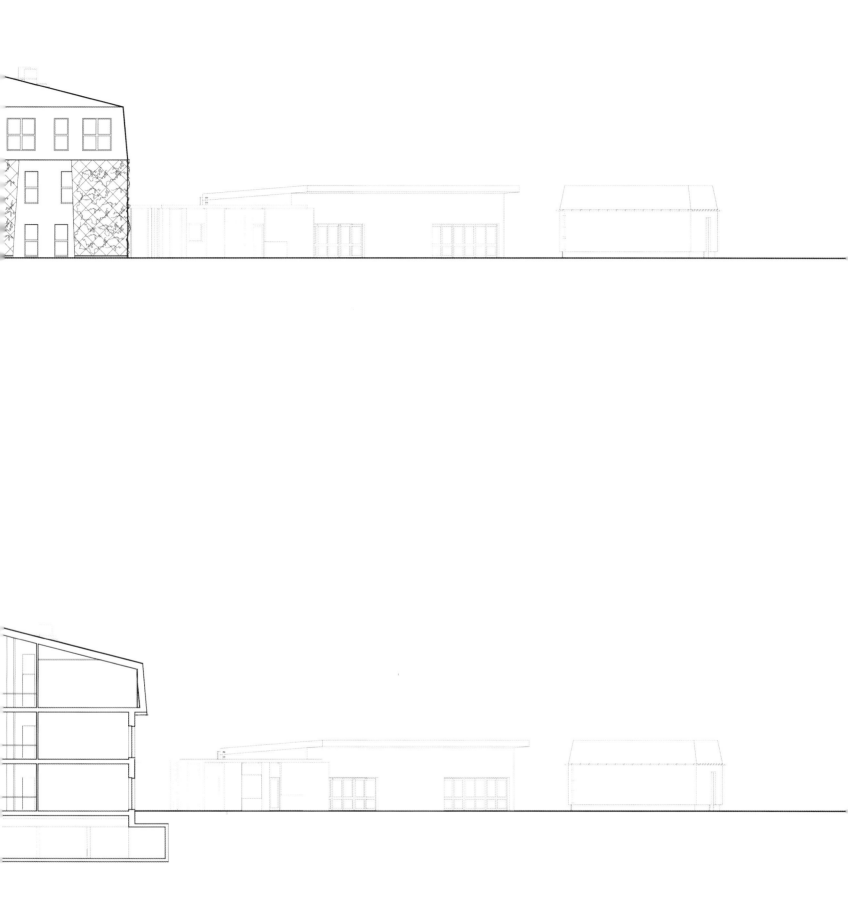

EVOLO摩天大楼
EVOLO

2013年
旧金山 / 美国
San Francisco / United States

摩天大楼竞赛是由美国建筑杂志EVOLO举办的,每年一次,旨在开发创新项目。建筑团队的想法是在经历生态灾难之后建立一个自治的城市——生活围绕着一座水坝展开,它可用于生产能源、灌溉庄稼,从而使自给自足的生活成为可能。出行主要依靠飞艇或船只,人们居住在带有小花园和游泳池的塔楼内。

这一竞赛赋予建筑团队在常规建筑思维之外开展项目的机会,让建筑团队能够思考备受关注的主题,如生态污染、自然资源稀缺、人口增长等多种社会问题。

整个公司通过头脑风暴的方式完成了该项目的设计,创建了一个乌托邦式的城市,并对其进行了深入的研究。

The Skyscraper contest is organised annually by the American architecture magazine EVOLO with the aim of creating an innovative project. Our idea was for an autonomous city after an ecological catastrophe. Life is organised around a hydraulic dam that makes it possible to produce energy, grow crops, and live self-sufficiently. Journeys are made by airship or boat. People live in towers with small gardens and swimming pools.

The Evolo Contest was a chance to work on a project that exists outside the usual architectural framework, and that allows us to think about key topics such as pollution, ecology, social problems, the scarcity of natural resources, population growth, and so on.

The entire firm participated in the project via brainstorming sessions. It allowed us to think about precise topics, to create a utopian project, and to carry out research.

罗斯福高中
Roosevelt High School

2015年
米卢斯 / 法国
Mulhouse / France

学校位于米卢斯市中心马尔凯广场和罗斯福总统大道之间，其原有建筑已经非常过时，无法满足现代人的需求了。

学校包括两座建筑，一座建于20世纪初期，一座扩建结构建于20世纪70年代。前者气势磅礴，以一种压倒性的姿态存在，而后者努力确保其"公共机构"的地位。

业主要求对高中大楼进行翻新，包括对建筑物的改造，也包括对整个校园的改造。

建筑团队将主入口移到庭院内，以确保进入学校的便利性。如今，其被扩建部分结构遮蔽，并与建于20世纪70年代的建筑相连。

扩建部分采用铜铝合金包层饰面，在庭院里闪闪发光，犹如一个小巧精致的稀有物。其优雅而俏皮的造型在新入口上方形成一个门廊，并固定在起到支撑作用的混凝土台阶内。在内部打造的一个垂直动线空间，成为学校新布局内的核心元素。

建于20世纪70年代的建筑外立面采用刀片状铜板覆盖，犹如波浪一般，使其与外界隔离开来。这一设计使建筑仿佛恢复了往日的辉煌，并重新构建了与相邻建筑的关系。铜板构筑出光滑的立面，伴随着行人一起穿过狭窄的人行道。

北立面采用落叶松木覆层打造，为庭院注入温馨的气息。竖边线使用镀金铜装饰，与新入口和南立面交相呼应，并突出了木覆层的水平形态。

In the city centre of Mulhouse, between the Place du Marché and the Boulevard du Président Roosevelt, the Lycée des Métiers no longer met modern expectations and was completely outdated.

The establishment is composed of 2 buildings, a well-constructed building from the beginning of the 20th century as well as an extension dating from the 1970s. The latter struggled to establish its position as a « public institution » because of the overwhelming presence of the 20th century building.

The studio was asked to carry out the renovation of the High School's building. This included updating and refurbishing the building as well as restructuring the entire school grounds. In order for the project to be more easily accessible, the main entrance was moved to the courtyard of the school. It is now sheltered by a small extension which is attached to the 1970s part of the school, where the two existing buildings meet.

This new shinny and golden extension, covered with a copper/aluminium alloy cladding, shimmers in the courtyard like a « small and precious object ». Its graceful and playful shape forms a porch above the new entrance and is anchored in the dense mass of the concrete steps that support it. The extension hosts a new vertical circulation system and became the central element of the school's new layout.

The 1970s building is insulated from the outside. The southern facade is covered with shades, made of wavy copper blades. This elegant design restores the building to its former glory and rebalances its relationship with the neighbouring structure. The fluid and horizontal aspect of the copper blades accompanies the pedestrians through the narrow Rue de la Promenade.

To the north, the facade is adorned with vertical larch cladding, which brings a more playful and a warmer atmosphere to the courtyard. Only the splays and mullions are covered in gold-plated copper, echoing the new entrance and the south facade while underlining the horizontality of the larch cladding.

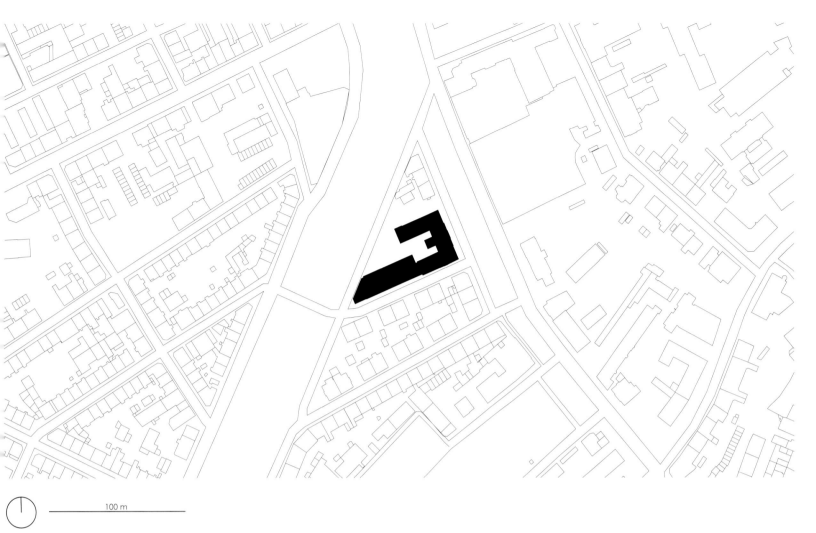

100 m

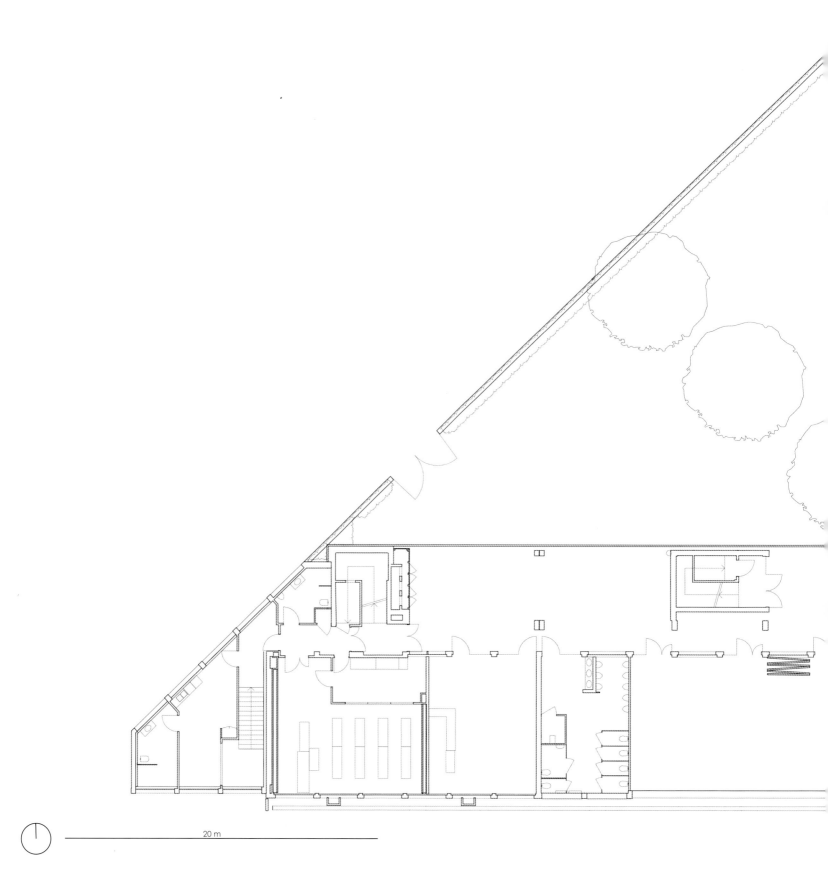

20 m

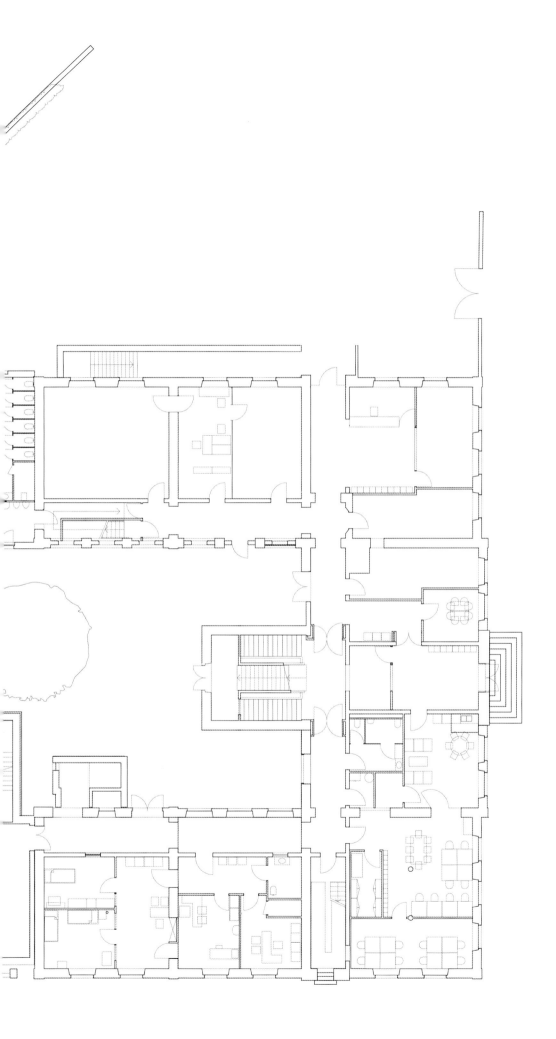

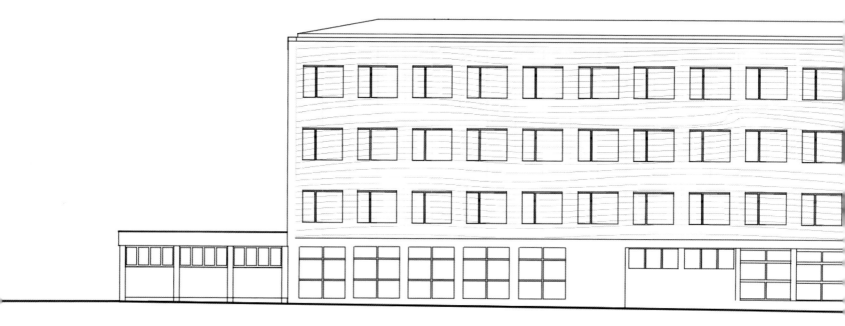

10 m

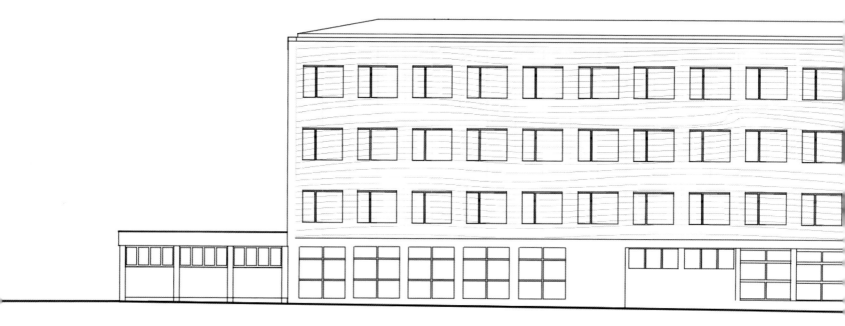

10 m

10 m

住宅、幼儿园和商店
Housing, Nursery and Shops

2016年
苏弗尔韦埃尔桑 / 法国
Souffelweyersheim /
France

感官之眼生态区包括住宅、幼儿园和小型商店，住宅楼四周环绕着郁郁葱葱的种植丛，住宅楼共包括157个居住单元。设计宗旨是将建筑融入周围景观之中，场地上原有的商业设施和垃圾场都被彻底拆除，并做了消毒处理。

设计团队对建筑物的位置和体量进行了仔细的思考，以尽量最小化其在欧洲大街和砖厂大街的存在，确保居民的隐私。每栋建筑都以简约紧凑的造型为特色，共为3层，性能优异且活力十足。阁楼从正面退后，确保从一层无法看到。

住宅及商店建筑沿南北轴线展开，朝向中央结构及绿地开放。幼儿园沿东西轴线设置，格外引人注目，清晰地彰显了其作为公共空间的功能。

整个建筑群内设有多条人行道，构筑了一个小型社区。两个室外停车场供幼儿园和商店使用。其余停车场设置在地下，共有3条坡道通向那里。停车场内设置人行道及绿地。

一条东西走向的人行道贯穿整个社区，其时而变宽，形成幼儿园前庭，其时而变窄，用于连通南北住区。此外，人行道两侧设计有种植带。

植被是整个项目的核心，每一栋建筑都被私人花园和种植树篱包围，提供必要的隐私，绿色植物与建筑的木质覆层及白色饰面相互衬托。

所有的住宅楼都带有宽敞的户外空间，或是一楼的私人花园，或是二楼和三楼的长露台，或是阁楼的屋顶。

Hidden behind lush planted clumps, the 8 buildings of the « l'Orée des sens » eco-district gather 157 apartments, a kindergarten, and a small grocery store. They are designed to blend into the residential landscape and the surrounding gardens. The business facility that used to occupy the plot as well as the former dump have disappeared. The latter required the site to be completely decontaminated.
The thought process behind the position and size of the buildings has helped to minimise their presence along the Rue de l'Europe and the Rue de la Briqueterie. This guarantees the peace and privacy of the residents. Each building has a simple and compact volume of three levels which helps to optimise its energetic performance. The attics are set back from the facade, so as to be invisible from the ground floor.
Positioned along a north-south axis, the buildings open up to the core of the project as well as the green areas. The kindergarten stands out and shows a more public face being the only building oriented along an east-west axis.
The project is designed as a small neighbourhood thanks to a network of pedestrian walkways that provide structure to the entire project. Two outdoor parking lots, used by the kindergarten and the small shops, can be seen in the neighbourhood. The rest of the parking lots are managed underground, with three access ramps punctuating the project. The management of the parking lots allows to create an exclusively pedestrian and green area.
An east-west pedestrian axis crosses the entire eco-neighbourhood and widens at times to create new public spaces: the forecourt of the kindergarten, a plaza... Narrower perpendicular paths connect the neighbourhood from the north to the south, whilst forming small planted lanes on both sides.
Vegetation is at the heart of the project and accompanies each architectural gesture. Each building is surrounded by private gardens and a planted hedge that gives it the necessary privacy. This greenery acts harmoniously with the wood cladding and white render of the new buildings.
All the apartments have large outdoor spaces, either in the form of private gardens on the ground floor or as long balconies on the second and third floors as well as rooftops for the attics.

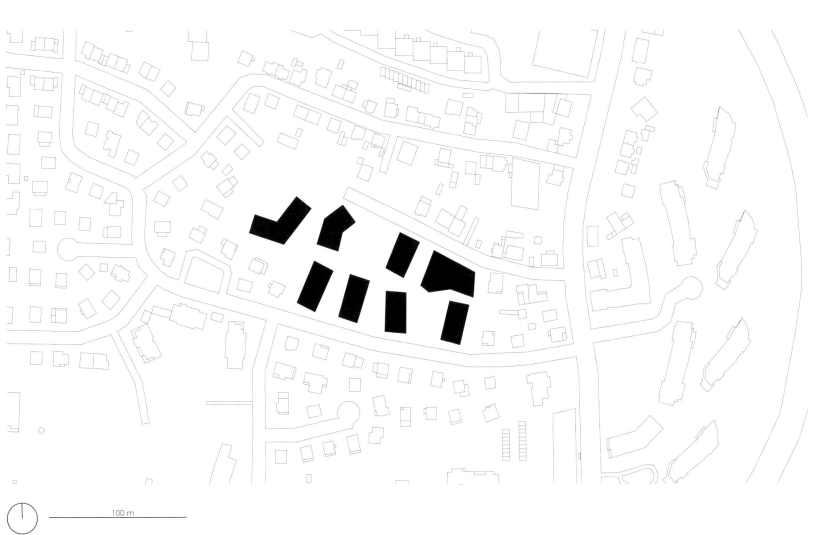

20 m

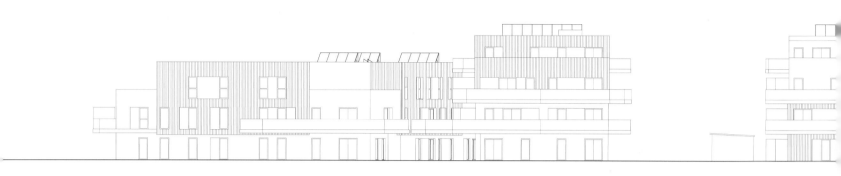

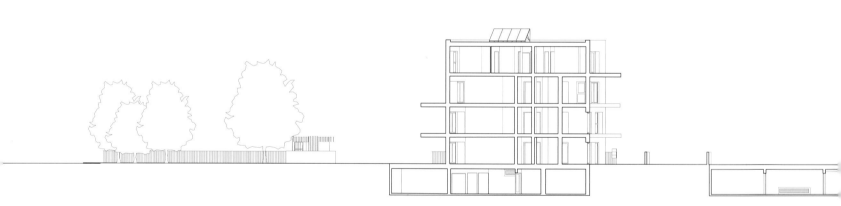

10 m

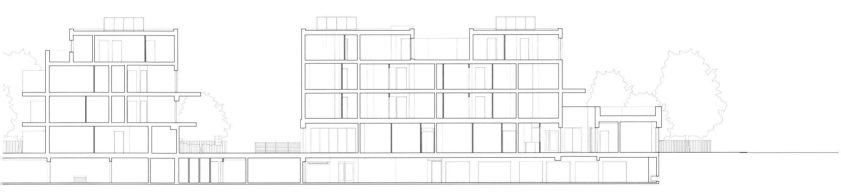

医院
Hospital

2019年
基加利 / 卢旺达
Kigali / Rwanda

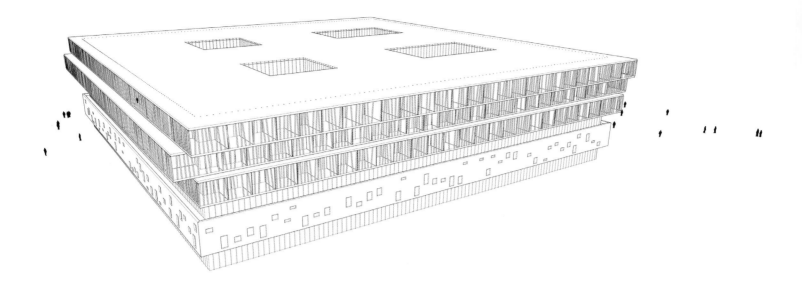

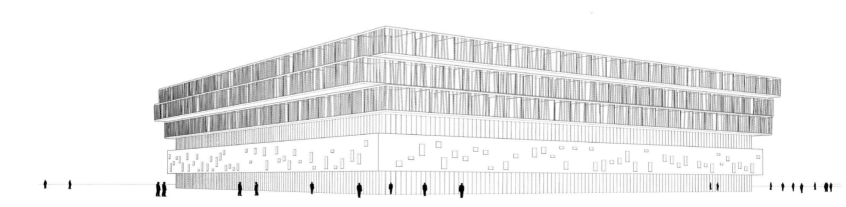

大学医院研究所
University Hospital Institute

2016年
斯特拉斯堡 / 法国
Strasbourg / France

全新的大学医院研究所位于斯特拉斯堡民用医院中心区——消化系统癌症研究所和新民用医院之间,主要致力于机器人引导微创手术的研究。其设计理念是植根于当下,着眼于未来,实现不同建筑风格的蓬勃发展与互动。

大学医院研究所内包含接待办公区、教学平台、研发平台、与新民用医院直接相连的临床平台以及与消化系统癌症研究所相连的实验平台。基于这个项目,建筑团队咨询了地标保护建筑师的意见,采用创新的动态设计方法,致力于展现建筑内部发生的各种活动。

整座建筑沿着新民用医院和消化系统癌症研究所原有的东西轴线(根据当地城市规划法规及周围建筑情况而设定)延展,体量简单。其呈现棱柱造型,犹如使用先进技术切割的钻石,缓缓盘旋上升,在东侧达到合理高度之后戛然而止,在西侧与新民用医院高度持平。

顶部采用巨大的折叠状屋顶覆盖,每个面都呈现出不同的造型。其独特的存在增强了与新民用医院屋顶景观的联系。

南立面嵌入一个垂直的缺口并采用绿墙覆盖,这里作为研究所的入口,格外引人注目。光线穿过这里可以照进建筑内部,一层空间做了退后处理,一条带顶棚的长廊一直通往接待大厅并与新民用医院连接。

整座建筑在周围环境中脱颖而出,铝制外壳会随着光线变化而变化。外墙和屋顶的植被让人看到勃勃生机,而彩虹色金属立面则与研究所内发生的高水平技术活动相得益彰。

内部空间设计以安抚患者为主要理念。镶木地板、墙壁上的木装饰和由回收的穆拉诺玻璃制成的大吊灯增添了"人性化"的气息。中庭将不同楼层和不同功能区连接起来,突出模块化和适应性。梁柱系统赋予空间及立面布局最大的自由度和灵活性,北立面开凿出宽大的固定框架凸窗,便于大型机械设备操作。

该建筑符合RT2012标准,满足了高效能要求。另外,这里配备了集中技术管理系统,方便患者。

The new University Hospital Institute (IHU) shines in the middle of Strasbourg's Civil Hospital and addresses the considerable development of robot-guided minimally invasive surgery. Located between the Digestive Cancer Research Institute (IRCAD) and the New Civil Hospital (NHC), the project looks towards the future while being rooted in what already exists and interacts with a hospital site where all types of architectural styles have flourished over time.

The IHU houses a multifaceted and technical programme: reception and office space, a teaching platform, a clinical platform operated by the NHC directly linked to it, a research and development platform, and finally an experimental platform connected to the IRCAD. For this project, subjected to the opinion of the Landmark Preservation Architect, the studio went for an innovative and dynamic design, echoing the activities taking place within.

The building consists of a simple volume, stretched along the east-west axis that connects the IRCAD to the NHC, sculpted by current urban planning regulations and the projection of surrounding buildings. This prismatic volume, a reference to the diamond used in cutting edge technologies, rises gently, spiralling upwards to stop at a reasonable height on the east side, then reaching the level of the NHC on the west side.

A large folding roof covers it and reveals several faces. Its remarkable presence reinforces the link with the old Civil Hospital's roofscape, and makes it possible to cover all the technical exits needed to operate the building.

A vertical gap is carved into the southern facade. Covered by a green wall, this gap catches the eye from the Quai Louis Pasteur and marks the entrance to the institute. On the inside, it provides light to the core of this thick building. Set back from the facade, the ground floor opens up a covered gallery that runs along the building, leading to the reception area and creating a pedestrian link with the NHC.

The IHU stands out without ostentation among the older buildings, thanks to its aluminium envelope that changes with the light. The vegetation on the facade and roof refers to the organic world and life, while the iridescent metal facade echoes the high level of technological knowledge and skill deployed within the Institute.

The interior is designed to soothe the patient. The parquet floor, the wood on the walls and the large chandelier made of recycled Murano glass give a more «human» image to the hospital. An atrium links the floors and their different functions with a focus on modularity and adaptability. The post-and-beam system allows great flexibility in the layout and freedom in the design of the facades. The north facade features a large fixed frame bay window to facilitate the handling of the necessary machinery.

The institute complies with the RT2012 standard and achieves high energy performance. The building incorporates a centralised technical management system (GTC) which allows consumption to be adapted to actual activities.

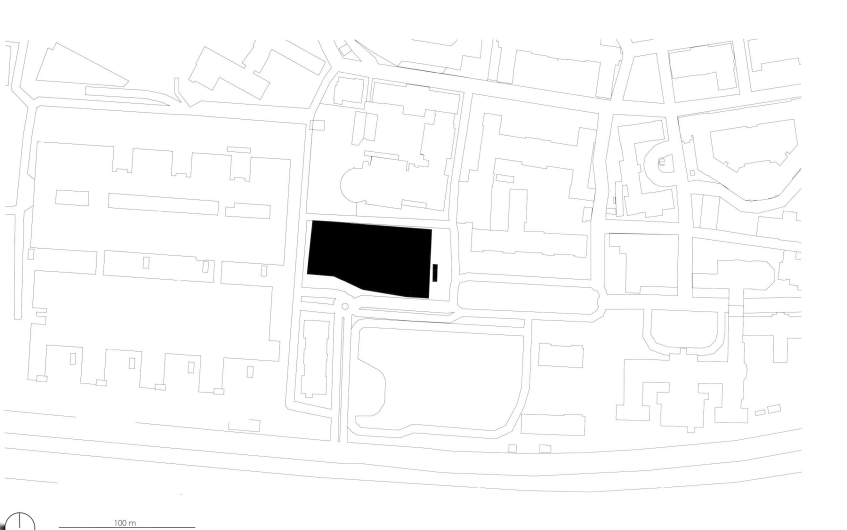

10 m

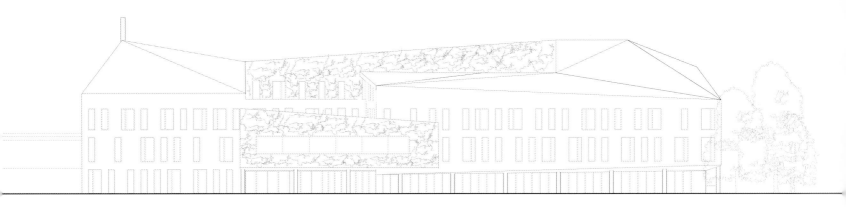

10 m

公司餐厅和仓库
Company Restaurant and Store

2018年
埃尔斯坦 / 法国
Erstein / France

福士集团计划在入口附近建造公司餐厅和仓库,并使其远离场地内的其他建筑。设计团队建议将其放在现有建筑物之间,以提高其可见性和可到达性,并构筑一个连贯的公司形象。

为此,新建筑选址在博物馆和另一侧宏伟的物流中心之间,成功地在两个地块之间建立了联系。其从道路上即可见,无疑成为整个场地的焦点与地标。建筑西立面在视觉上定义了博物馆和总部大楼之间的绿地,并突出了绿地作为接待区的主要功能。

整座建筑沿东西轴线伸展,连接物流中心和博物馆。一条宽阔的带顶棚走廊将建筑纵向分开——北侧是可以从乔治·贝斯大街看到并进入的仓库,南侧是餐厅以及阳光明媚的露台。

该建筑在风格上纯粹而现代,是对总部大楼和博物馆的完美回应——清水混凝土板外墙与博物馆外观交相呼应。在这里,建筑团队

选择混凝土作为主要原料,除了其维护成本低之外,更能凸显建筑本身的重要性。混凝土赋予建筑高贵的气质,并增强了照明效果。

商店外墙采用红色金属框架打造,具有浓郁的工业化风格,在整体的水泥外观之中格外突出,并与集团特有的红色标志相互呼应。

新建筑四周的大柱子暗示着集团总部大楼的高度。

一块巨大的悬垂混凝土板将餐厅、厨房和仓库3个部分整合在一起,可以起到保护外墙的作用,下面的构筑空间可作为室外露台和人行道。

餐厅内部采用了完全不同的材质——木材打造,如木质家具和木顶棚。整个空间散发出温馨友好的气息,其布局规划以灵活为主要理念,移动的家具和隔板可以组合出多个小空间。

The Würth Group planned to build its company restaurant and warehouse near the entrance of the property, away from the buildings scattered around the plot. Instead of that, the S&AA studio suggested placing it in the midst of the existing buildings so as to improve its accessibility, enhance its visibility and create a coherent design.
The new building is therefore built between the museum and the imposing logistics hub, on the other side of the Rue Georges Besse, and creates a link between the two plots. Highly visible from the road, the building becomes the focus point of the site, making it an important landmark. The western facade of the project visually defines the green space between the museum and the headquarters and emphasises its function as a reception area.
The facility stretches along the east-west axis connecting the logistics platform to the museum. A broad covered street splits the building lengthwise and separates, to the north, the shop which is visible and accessible from the Rue Georges Besse, and, to the south, the restaurant and its sunny terrace.
This corridor covers the flows of the logistics platform as well as those of the museum and the main office.
The architecture of the new building, pure and modern, is an answer to the architecture of the headquarters and the

museum. The facades, made of raw concrete panels, echo the museum's appearance. The choice of concrete as the main material for the project, in addition to being low-maintenance, reinforces the importance of the building. The material gives the building a noble character and enhances the lighting effects generated by its various volumes.
The shop's facade, with its red metallic frames and its industrial design, is surprising in the midst of this concrete architecture, and symbolises the Würth company, its activity and refers to its red logo.
The columns that surround the building hint at the high pillars of the headquarters' building.
A large overhanging concrete slab covers the three parts of the project - the dining room, the kitchen and the warehouse - and unifies the project as a whole. It protects and draws the exterior spaces, which then assume specific functions: terraces and walkways.
Inside, the dining room, designed as a large friendly space, offers a completely different materiality; wood is very present - in the furniture and on the ceiling. It fills the room with a warm and safe atmosphere. The room's layout is designed to guarantee its flexibility: it is possible to create numerous sub-spaces thanks to the furniture or the mobile partitions.

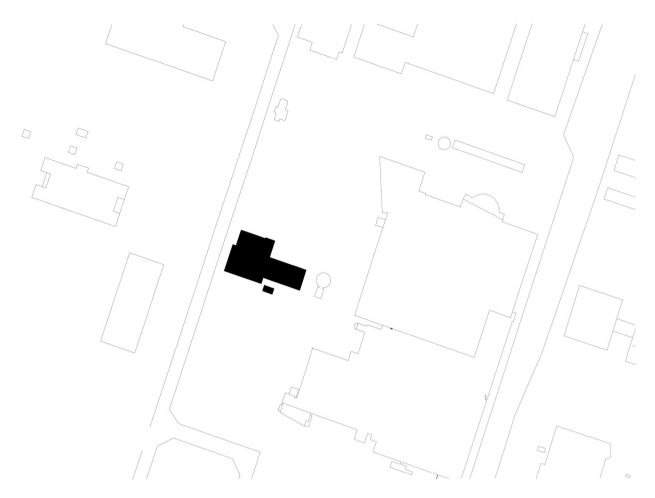

100 m

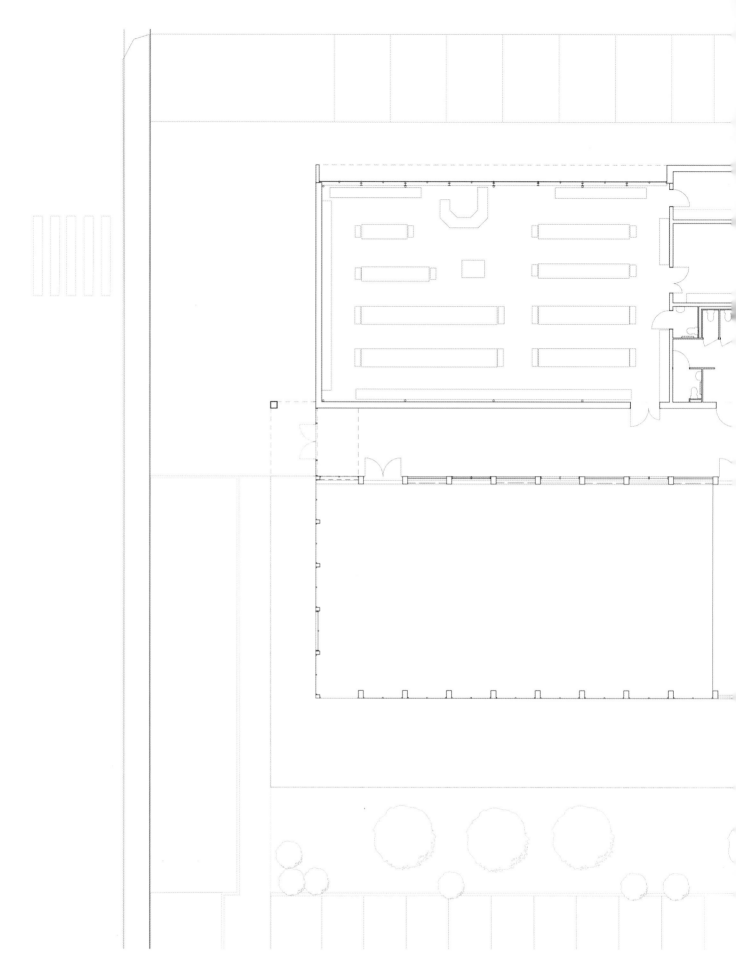

10 m

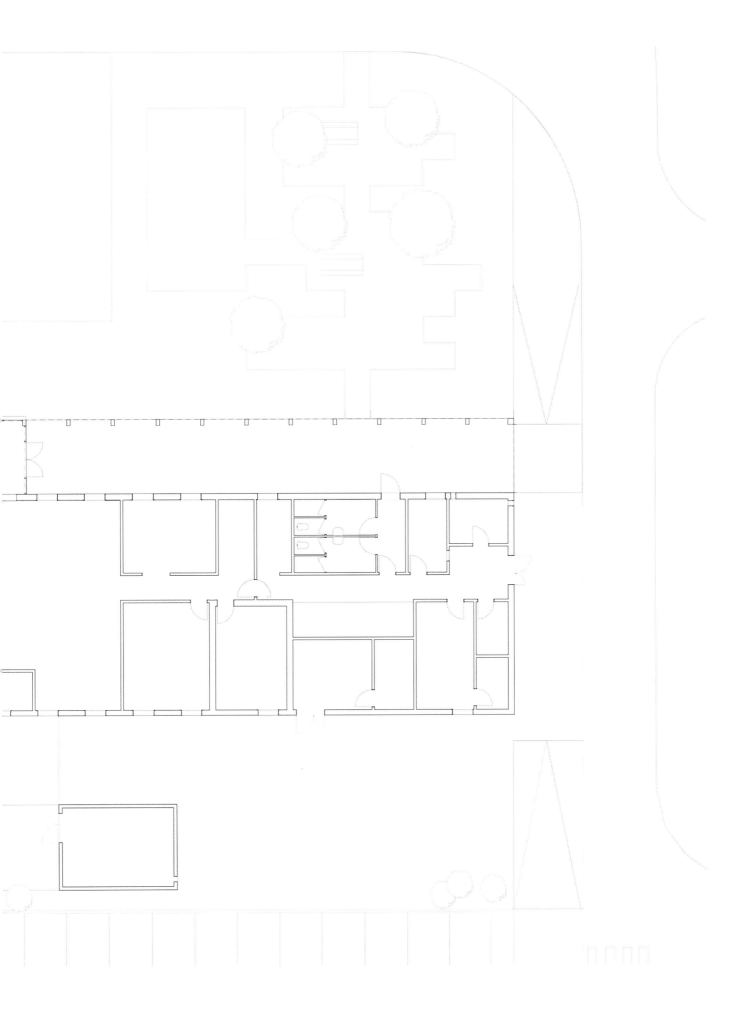

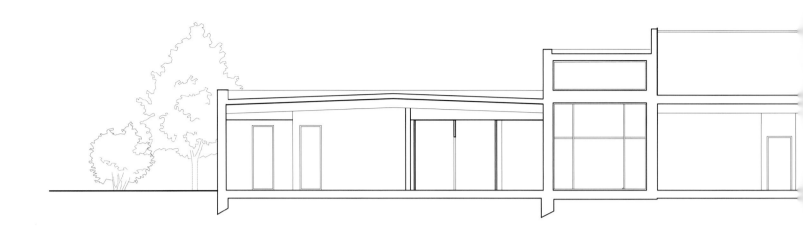

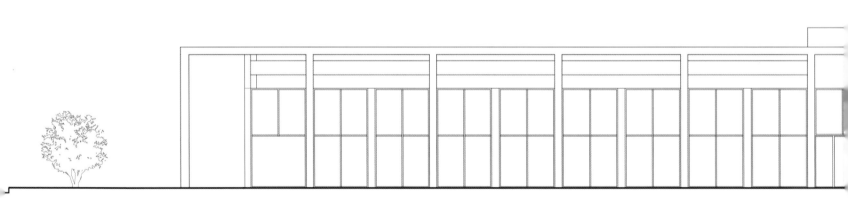

10 m

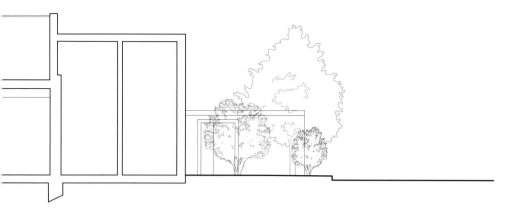

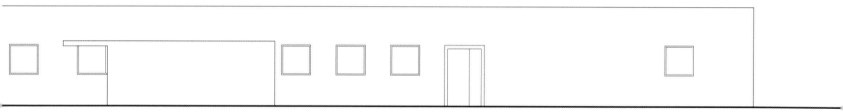

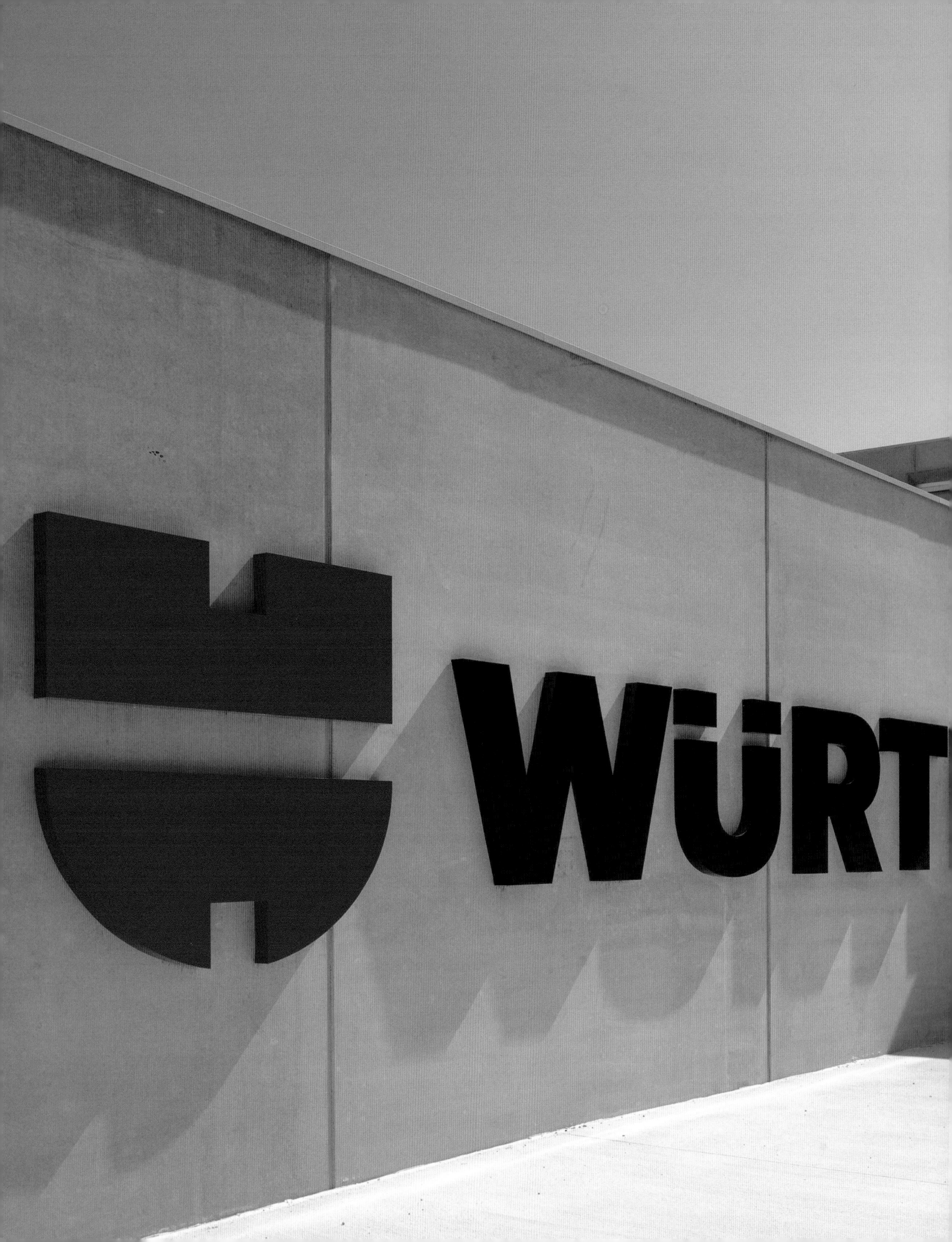

七个机场
Seven Airports

2019年
登比多洛、戈德、金卡、孔博勒查、内格默特、罗宾、希雷 / 埃塞俄比亚
Dembidolo, Gode, Jinka, Kombolcha, Nekemte, Robe, Shire /
Ethiopia

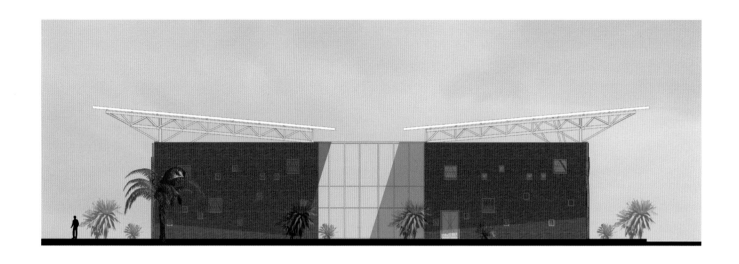

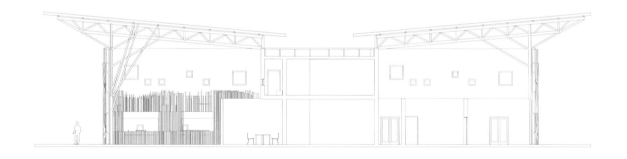

旧档案馆修复
Rehabilitation of Old Archives

2018年
斯特拉斯堡 / 法国
Strasbourg / France

下莱茵省档案馆大楼是德国政府在阿尔萨斯–摩泽尔帝国直辖领建造的唯一一座档案馆,其选址在新城社区中心,因独特的纪念性而格外引人注目。整体包含3座建筑,始建于1890年,拥有高雅的新文艺复兴和新古典主义风格。21世纪中期,档案馆被迁走之后,这里就空置下来,也开始了逐步的修复。

如今,昔日的档案馆大楼被改造成容纳42套公寓的精美住宅建筑,焕然一新。原有建筑外墙和主体金属框架以及室外护栏和铁艺制品被保留下来并进行了翻新。旧楼梯被拆除之后重新安装到不同的楼层。3座建筑位于菲沙特大街的主入口,经过了修缮,主入口被打造成小型的装饰花园,别有一番特色。原有的两个两层高的画廊依然用于连接3座建筑,但一层被改造成了居民休闲生活空间,可以直接进入庭院一侧的私人花园。

进入位于孚日大道和菲沙特大街拐角处的建筑之后,可以直接通往壮丽的入口大厅,这里保留了原有的旧楼梯,通向不同的公寓住宅。新旧的完美融合代表着高雅、舒适的风格以及建筑遗产的传承。

一座曲线造型的新建筑矗立在庭院中,隐藏于种植墙后,从大街上几乎不可见。粉红色的砂岩外立面毫无疑问能够给人留下非常深刻的印象。

其醒目的形状是为了顺应场地的多面性,希望尽可能多地保留原有建筑物内外的景致,并为庭院花园带来更多的光线。

其分层结构设计让14个新建的公寓住宅能够拥有超大露台。这些露台的造型随着建筑立面的蜿蜒而延展。

弯曲的立面部分与庭院背景周围的建筑融为一体。

设计团队特意将新建筑与旧档案馆大楼区分开来,并将旧建筑视作庭院内的"雕塑"。新公寓现代而明亮,与由原有建筑改造而成的高品质公寓相互呼应。

建筑沿孚日大道一侧立面采用耐候钢材质打造,似乎在保护着街区中心的入口以及通往地下停车场的坡道。

As the only archive building built by the German administration in the Reichsland of Alsace-Moselle, in the heart of the Neustadt neighbourhood, the building of the Bas-Rhin's departmental archives stands out by its monumentality. The facility is made up of three buildings, which were built from 1890 onwards in an elegant neo-Renaissance and neo-classical style. In the mid-2000s, the departmental archives moved to new premises. The building was left empty and its restoration process could begin.
Today, the former archives have been given a new life and have been transformed into 42 beautifully designed flats. The shell and the metal structure of the buildings were completely preserved and renovated, as well as the exterior frames and the ironwork. The old staircases were removed and reinstalled in the various flats. All three buildings maintain their entrances on the Rue Fishart, which are now highlighted by small ornamental gardens. Both two levelled galleries connecting the buildings with each other, are turned into living spaces for the residents and allow, on the ground floor, access to the private gardens landscaped on the courtyard side.
The building, at the corner of the Avenue des Vosges and Rue Fishart, leads into a superb full-height entrance hall in which the old staircase has been preserved, and which now leads to the various flats. The old and the new combine on behalf of elegance, comfort and architectural heritage.
Barely visible from the street, hidden under a planted facade, a building with curved lines stands in the courtyard at the bottom of the impressive pink sandstone facades.
The striking shape of the building is the result of the numerous prospects on the site, the desire to preserve as many views as possible from and to the existing buildings, as well as the desire to bring light to the courtyard gardens.
The layered structure of the building allows the 14 new flats to have large terraces that follow the fluid movement of the facades.
The curved facades are partly clad in red brick to blend in with the surrounding buildings in the background of the courtyard.
The new building deliberately distinguishes itself from the old archive and emerges as a sculpture in the garden. The new apartments, on the other hand, bright and modern, match the quality of the flat designed in the existing buildings.
From the Avenue des Vosges, a Corten steel facade guards the entrance to the heart of the block as well as a ramp leading to the underground parking area.

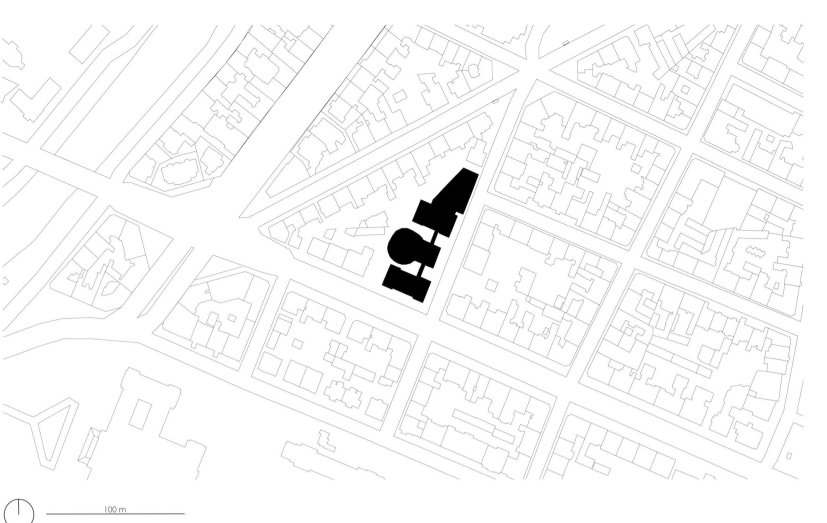

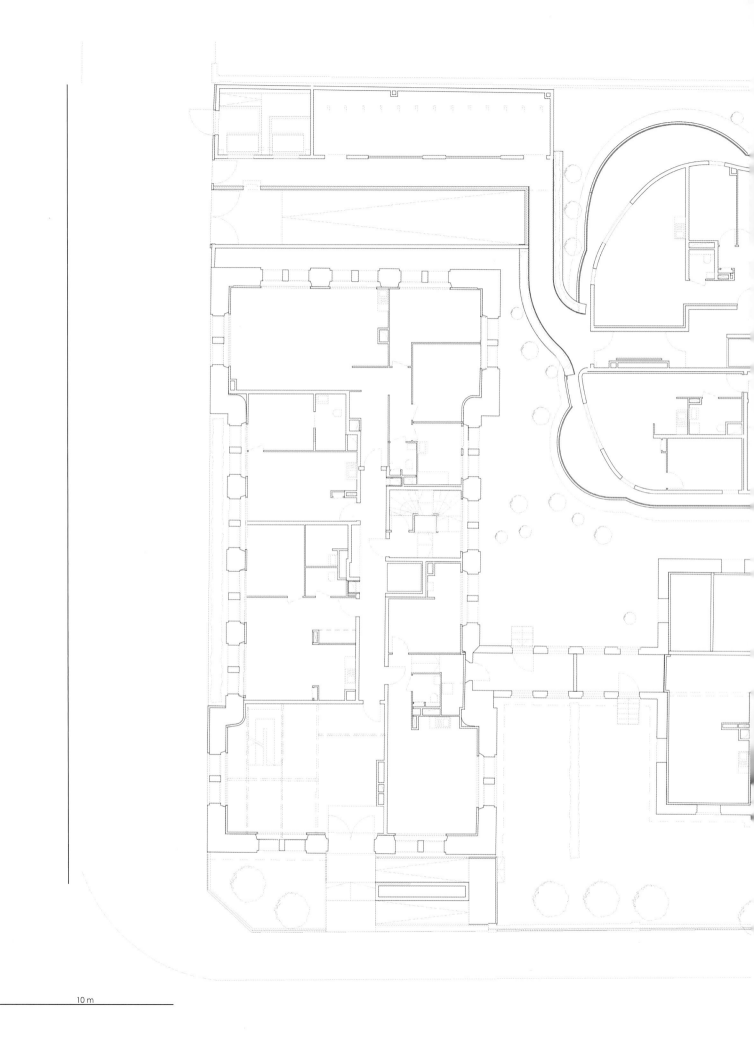

10 m

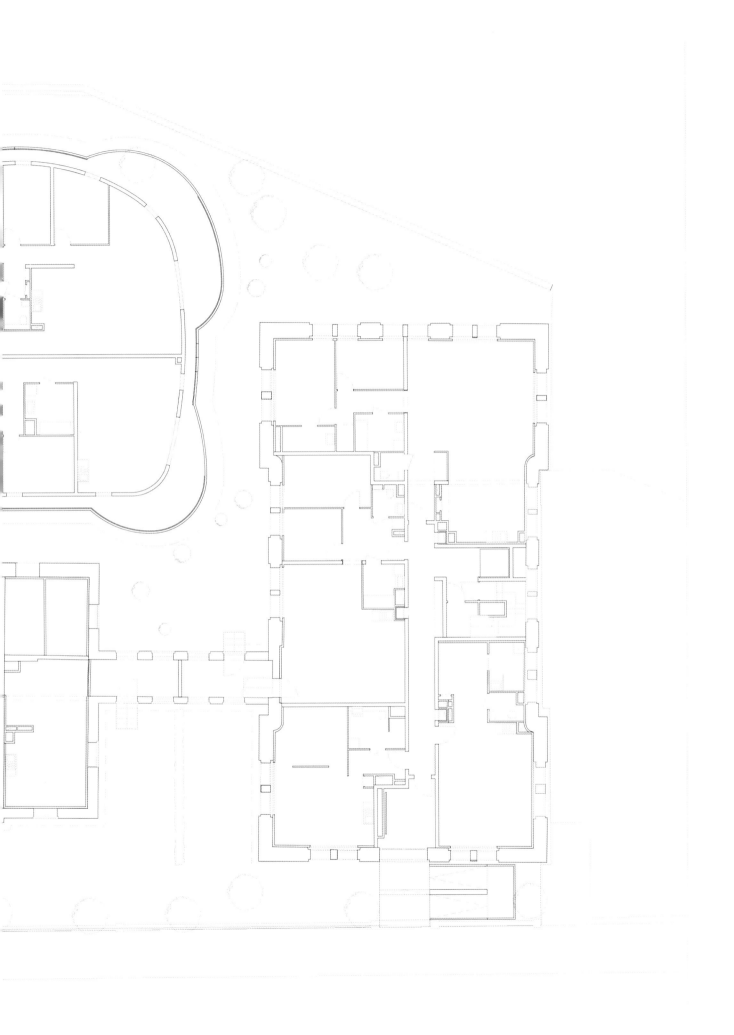

10 m

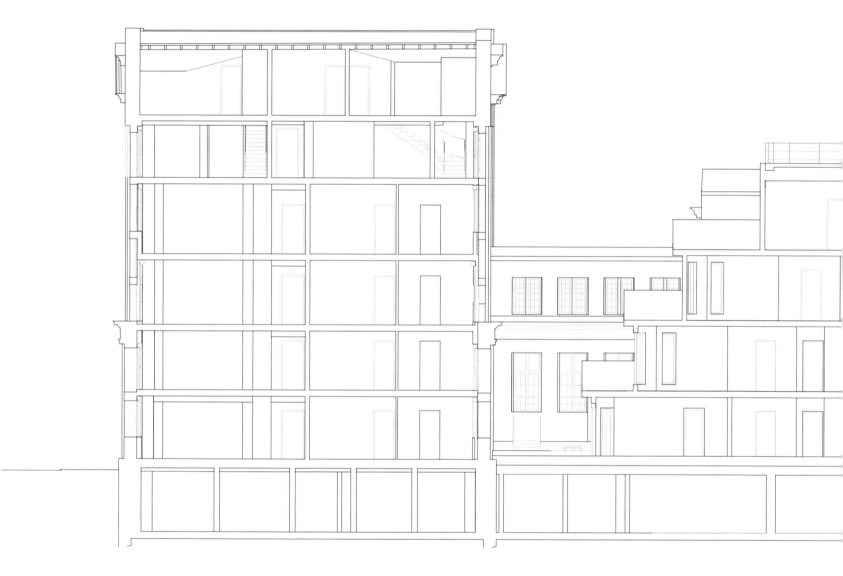

10 m

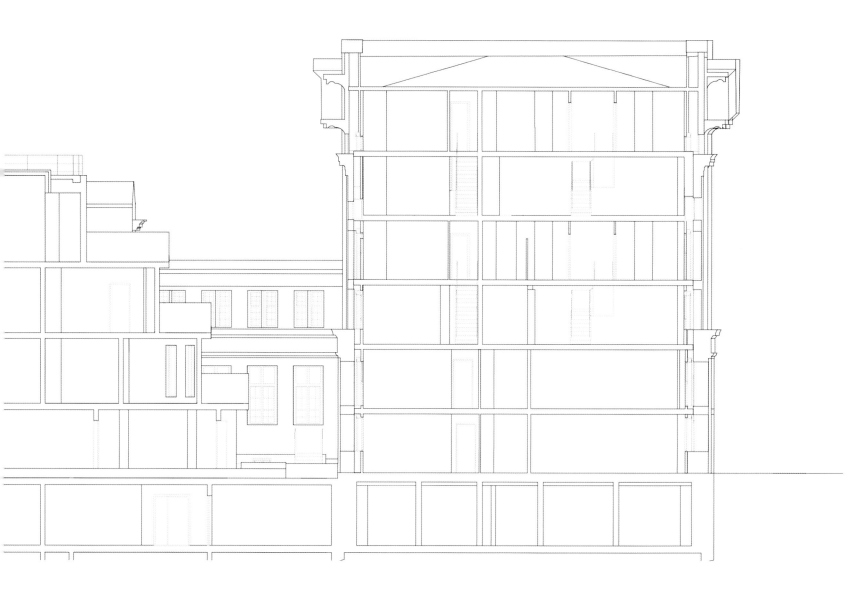

棕地修复
Conversion of a Brownfield Site

2018年
希尔蒂盖姆 / 法国
Schiltigheim / France

随着位于希尔蒂盖姆南部的Quiri锅炉制造厂的改造工程的实施，场地上的其他已有建筑的未来也被提上了日程，尤其是工业风浓郁的主楼。设计团队决定保留其巨大的金属框架，并将其作为整体项目的核心。

项目入口临近查瑟尔大街，带有壮观山墙结构的工业风大厅格外引人注目，但由于过于破旧而无法保留下来。为此，设计团队将墙壁完全拆除并采用相同的形式进行重建。

原有的铁艺结构也被完全移除，经过修复翻新依然安装在原来的位置上。而其他建筑太过老旧过时，且缺乏吸引人的元素，彻底被拆除了。

5栋新建筑沿东西向人行道排列，共包括107套住宅。主体建筑共为4层，造型简单——白色体块坐落于金属框架内，从屋顶处可以清晰看到它的构成。巧妙的构思打造了独特而生动的居住环境。另外4栋建筑用作集合住宅，造型简约，砖立面使其完美地融入周围的工业化氛围中。

两层的地下空间满足了停车需求并实现了棕地修复的可持续发展理念。位于一层的住宅带有南向花园，这是由原有工业大厅的高大砖梁改造而来的。

该项目是与地标保护建筑师合作完成的，已经成了棕地修复工程的标杆，并获得了许多奖项，如"2019年城市建筑设计奖""2019年设计灵感奖玫瑰金奖""2019年金字塔银奖"以及"2020年德国设计大奖卓越建筑奖"。

With the conversion of the former Quiri boilermaking factory in the south of Schiltigheim, arose the question of the future of the existing buildings, in particular the main building which is a true industrial cathedral. Under the spell of the old factory hall, the studio decided to keep this enormous metal frame and place it at the heart of the project.

By the Rue des Chasseurs, the entrance of the project is marked by the impressive gable wall of the industrial hall. Too worn out to be preserved, the wall was demolished and rebuilt identically.

The iron structure was completely removed, restored and reinstalled in its original location. The other existing buildings, which were outdated and less attractive architecturally speaking, were demolished.

The 5 new buildings consist of 107 dwellings and are organised along a new east-west footpath. The main building of the complex, a simple 4-storey volume, is built under the metal frame. Its white blocks are nestled under the skeleton of the structure which is visible from the rooftops that cover the duplex apartments. From the merging of these entities, new unique and distinctive living environments come to life.

The other four buildings are meant for collective housing. With their simple shapes and brick cladding they totally blend in with the industrial surroundings.

A two-levelled basement provides the required parking spaces whilst maintaining green spaces on the brownfield. The apartments on the ground floor benefit from south-facing gardens which are shaped by the tall brick pillars of the former hall.

The project, which was carried out in cooperation with the Landmark Conservation Architect (Architecte des Bâtiments de France), has become a reference in the field of brownfield requalification and has received several awards:
- The Urban Design Architecture Award in 2019
- The Muse Design Award in 2019
- The Pyramide d'Argent in 2019
- The German Design Award in 2020

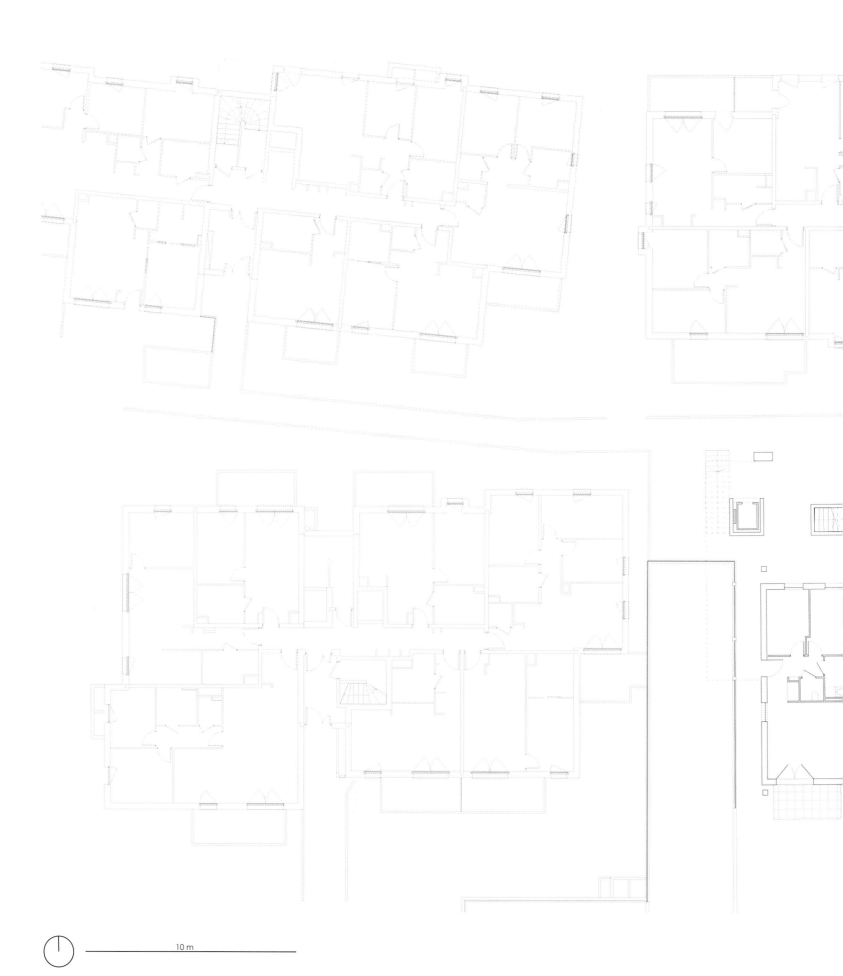

10 m

240

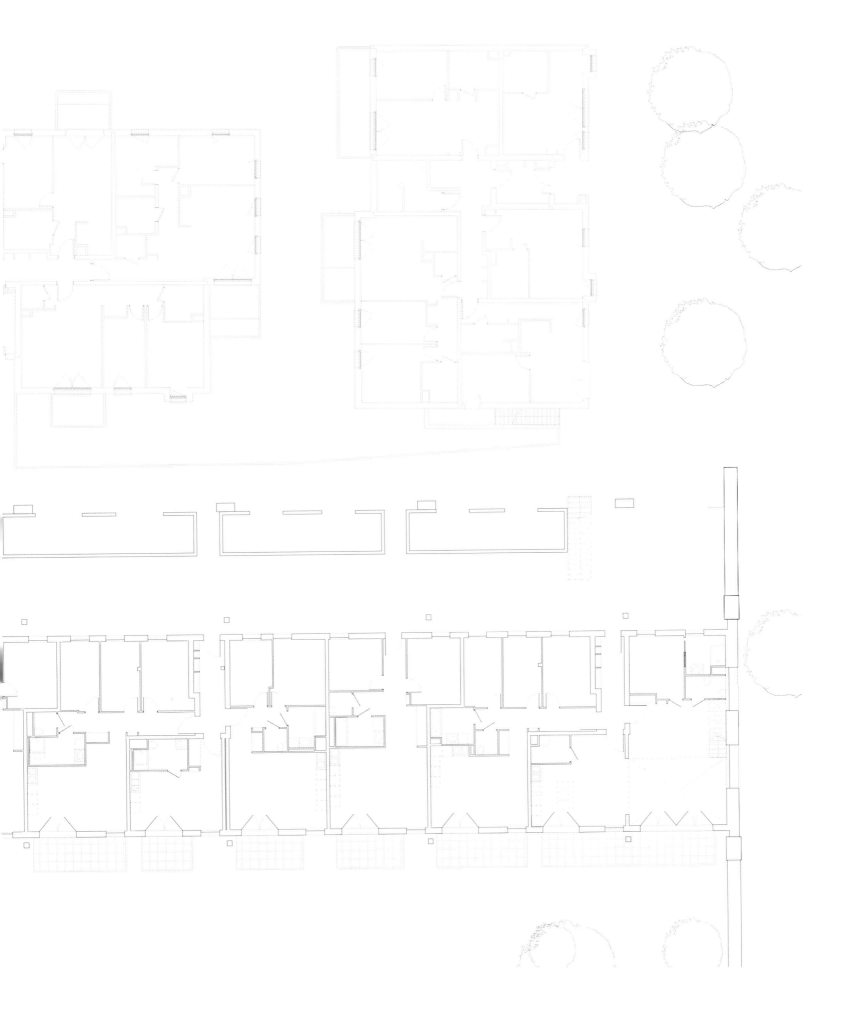

241

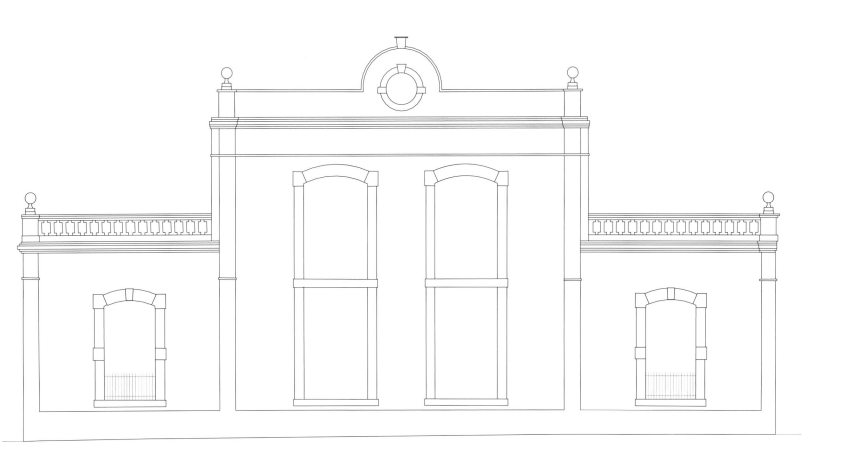

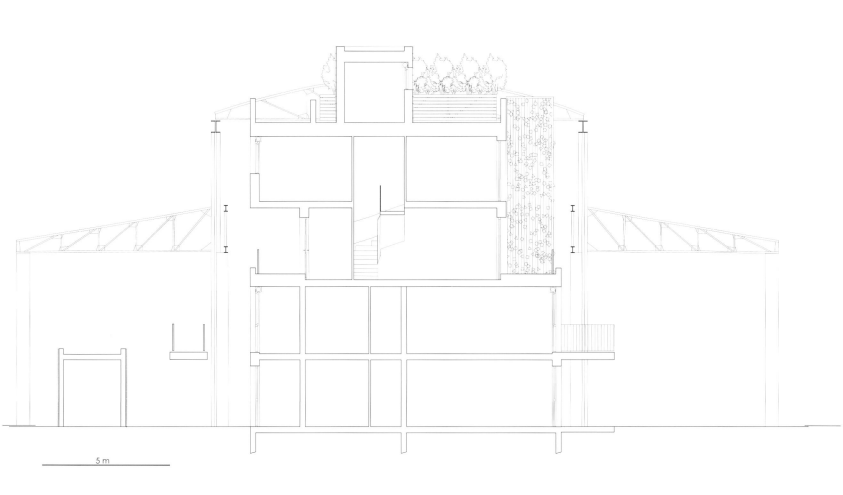

5 m

大邱孤山公共图书馆
Daegu Gosan Public Library

2012年
大邱 / 韩国
Daegu / South Korea

这座图书馆位于韩国大邱孤山附近,犹如雕塑一般置身于景观之中。该建筑四周是中型公寓楼,因此被打造成相似的体量,已与周围的环境融为一体。它的独特之处体现在带有锐角的造型上,立面由玻璃和实心穿孔金属板覆盖,可以映射出附近的环境。

北立面向公园倾斜,两者相映成趣。坡道使建筑得以延伸,最终创建了一个环形的公共空间,作为建筑本身的绿色庇护所。

大厅顶棚较低,里面汇集着各种终端设备、接待台以及报刊区。所有这些空间围绕着一直通往阅读区的大型楼梯展开。

来访者要先穿过不透明的储藏室才能到达开放阅读区。储藏室在这里被用作低矮的入口与宽敞、通透而明亮的开放阅读区之间的过渡。

屋顶映射着立面的倒影,将光影引入开放阅读区内。

The library in the Gosan neighbourhood of Daegu, in South Korea, was designed as a sculptural form in the landscape. Surrounded by medium-sized apartment buildings, the building blends into its environment by adopting a similar scale. Its uniqueness is expressed in its monolithic shape with sharp angles. The façade is composed of a multitude of triangles in glass or solid and perforated metal that reflect the environment.

The northern façade slopes towards the park and mirrors it. A ramp extends it to create a public space that wraps around the building. This space serves as a soft, green shelter for the angular building.

The hall, with its low ceiling, is a space that gathers the research terminals, the reception desk and a periodicals area, organised around a monumental staircase that leads to the reading areas.

Before entering the open space reading areas, the visitor crosses the opaque storage floor. This floor, close to the collections, serves as a transition between the horizontal, low-ceilinged entrance area and the vertical, aerial, transparent, light-filled reading areas.

The facade system is turned upside down on the roof, filtering the sun and creating shadows in the reading spaces.

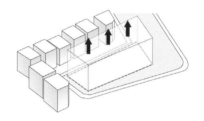
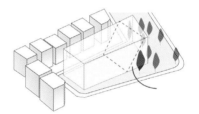
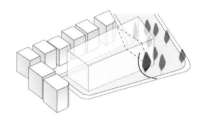

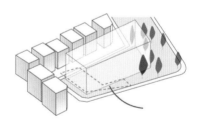
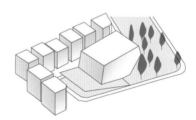
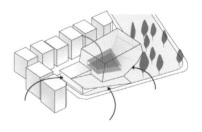

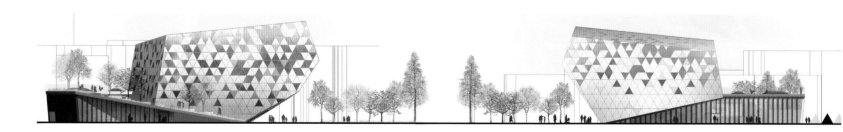

建筑与环境设计学院
Faculty of Architecture and Environmental Design

2018年
基加利 / 卢旺达
Kigali / Rwanda

新建筑与环境设计学院选址在基加利科学技术学院所在的山地上，距原建筑与环境设计学院不远，共可容纳600名学生。基于这一项目，设计团队采用了强劲的建筑学方法打造，运用复杂的形式诠释教学和环境问题。设计的宗旨是打造简单的教育环境，为快速发展的非洲国家培育未来建筑师。

整个项目的实施过程是综合考虑了建筑与地形关系的结果。设计团队参考火山形态，构思了棱柱造型，然后将其解构重组，如同在对抗来自外界的破坏力，从而赋予建筑强大的张力感。

设计团队充分利用不同结构之间的空隙，将其打造成室外活动空间，朝向基加利科学技术学院、山谷与城市，主要供学生们休闲使用。

建筑通过4种自然元素再现卢旺达景观，火山石和泥代表土，橙色代表火，动线空间指代空气，室内花园映射水。

该项目共为两层，总面积达5600m²。一层主要提供后勤功能，包括行政区、实验室、工作室、研讨室和礼堂。二层由13个不规则的棱柱结构构成，分别用作工作室、教室及展厅等。每个房间的体积都不尽相同，视角及景观也各有差异。

室外被视作整体空间，专门打造的座椅和台阶鼓励学生们之间的互动与交流。步行道将内部道路两侧的两座建筑连接起来，赋予空间足够的动线区域，并增强了活力。

设计团队打造了一栋本身具有教育意义的建筑，向学生诠释建筑的行为，这对于未来建筑师来说是非常重要的。他们用一种巧妙的方式鼓励学生利用当地资源，打造具有多种视角和规格的建筑，并使其与周围环境交相呼应。

该项目充分使用了当地资源，如火山石、原土、混凝土和木材，推崇简单、持久且易于维护的技术。这里没有电梯，仅有入口坡道；这里使用高效的自然通风系统和室内气候控制系统取代空调和暖气；这里的开窗设计旨在更多地引入自然光线，尽可能少地使用人工照明。混凝土墙具有良好的隔热和防水效果，尤其在外面刷上涂层之后，可以有效避免热量存储与释放。地面采用中空设置，便于收集雨水。

整个项目建设过程仅用了一年时间，致力于宣传当地传统知识与当地产业。400名建筑专业人士现场施工，铁匠和木工通力合作，地面是在现场浇筑完成的。

The new school of architecture, with a capacity of 600 students, is located on the same hill as the KIST (Kigali Institute of Science and Technology) not far from the Faculty of Architecture and Environmental Design, in Kigali. For this project, the practice has chosen a strong architectural approach that uses complex forms, driven by both pedagogical and environmental concerns. This is a simple and didactic design aimed at training the future architects of a rapidly developing African country.
The building is the result of a thought process taking architecture and territory into account. In reference to the volcanoes, the studio designed prisms which were then deconstructed, in the same way that tectonic forces distort land masses, creating areas of tension similar to canyons.
A central crack is formed creating an outdoor living space for students, which opens towards the entrance of the KIST, towards the valley, and towards the city.
The shapes and colours of the Rwandan landscape are represented in the building through the theme of the 4 natural elements: for earth, lava stone and raw earth; for fire, the colour orange; for air, the circulation areas; and for water, the inner gardens.
The entire 5600 m² programme is spread over two levels. The ground floor accommodates the school's logistics and facilities such as the administration, laboratories, workshops, seminar rooms and the auditorium. On the first floor, thirteen irregular prisms contain studios, classrooms and an exhibition gallery. Each room has a different volume, perspective and

view.
The outside is treated as a space in its own right, where seats and steps encourage interaction and encounters. Walkways link the two buildings on either side of the inner road. They allow the spatial articulation of the design and give a dynamic aspect to the project.
The studio decided to make a building that was educational in itself. It would be showing students the act of building, which is fundamental for a future architect, by encouraging them to use local resources and by creating various environments with multiple perspectives and volumes.
The building promotes the use of local resources - volcanic stone, raw earth, concrete, and local wood - and encourages the use of simple, lasting and easy to care for techniques. There is no lift, but there is an access ramp. Efficient natural ventilation systems and indoor climate control systems are used instead of air conditioning and heating. Openings designed to capture natural light are preferred to artificial lighting. The concrete walls are insulated, waterproofed and coated from the outside to avoid the storage and release of heat. The floor is marked by « cavities « in order to collect rainwater.
The construction process lasted only one year and sought to promote local and traditional knowledge as well as local industries. On site, 400 building professionals were summoned, locksmiths and carpenters set their workshops up, the floors were poured on site.

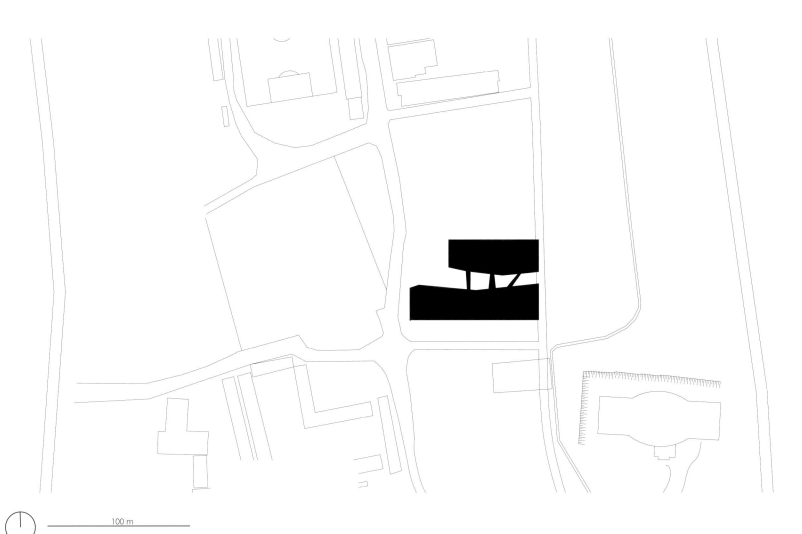

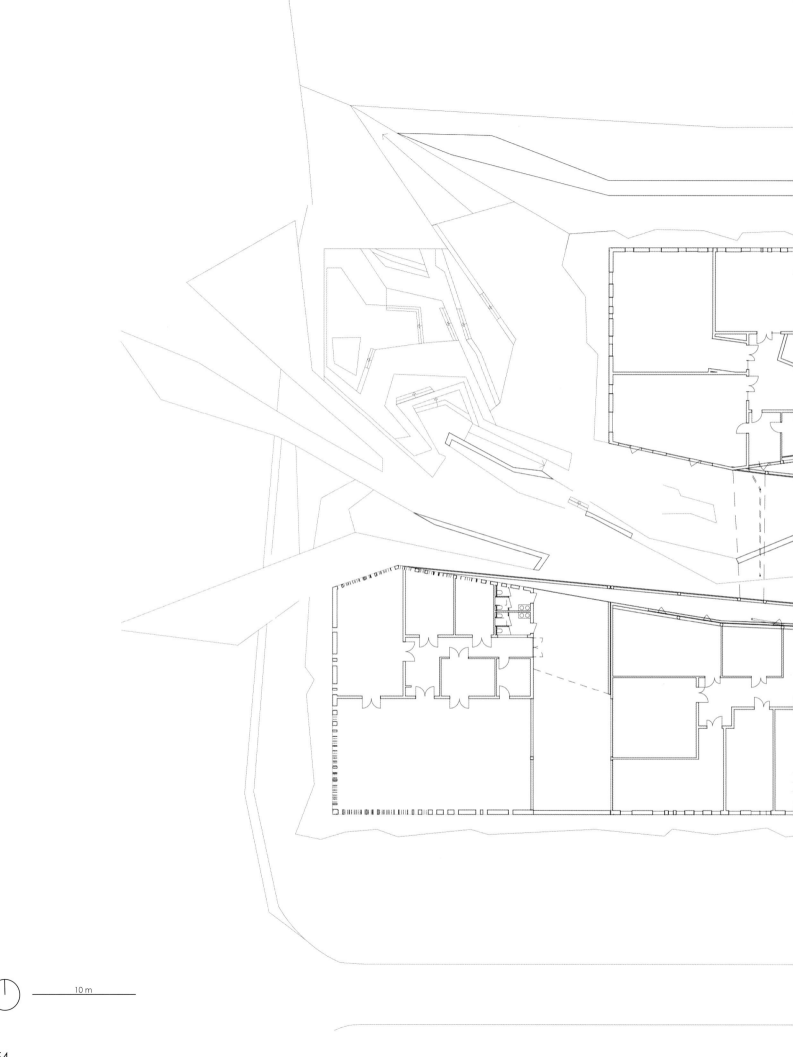

10 m

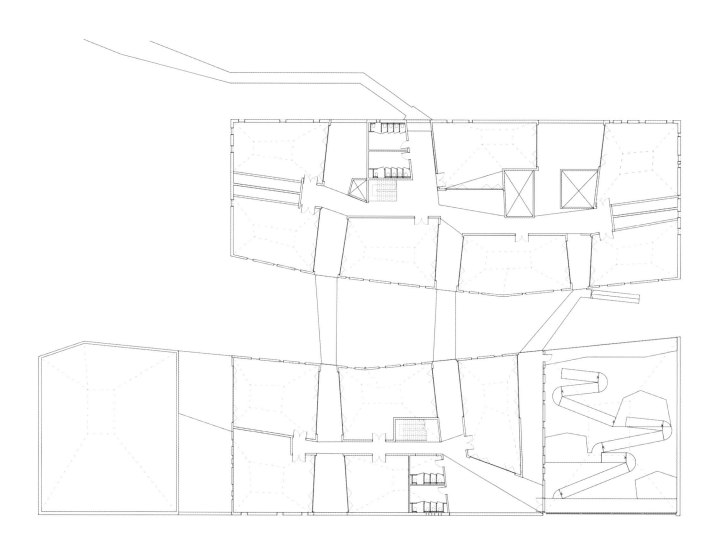

10 m

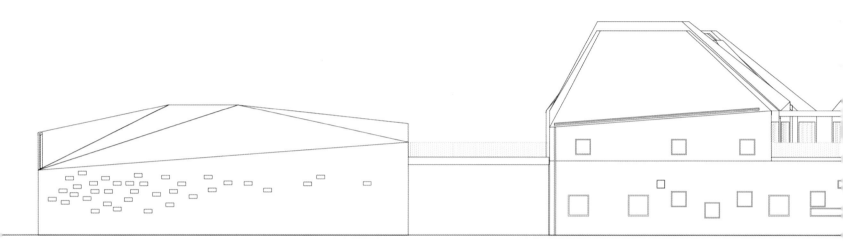

20 m

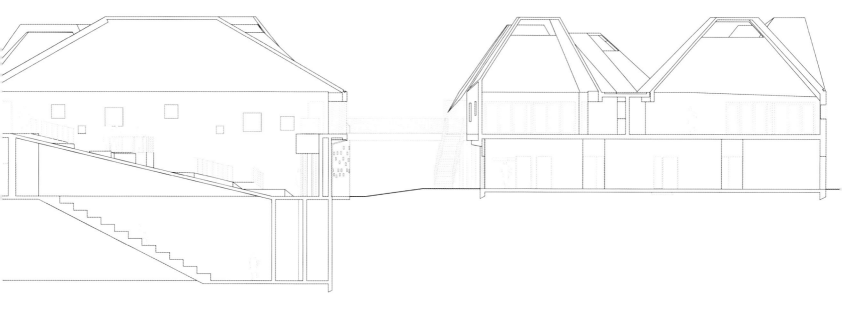

10 m

中国消化系统癌症研究所
IRCAD China

2019年

无锡 / 中国

Wuxi / China

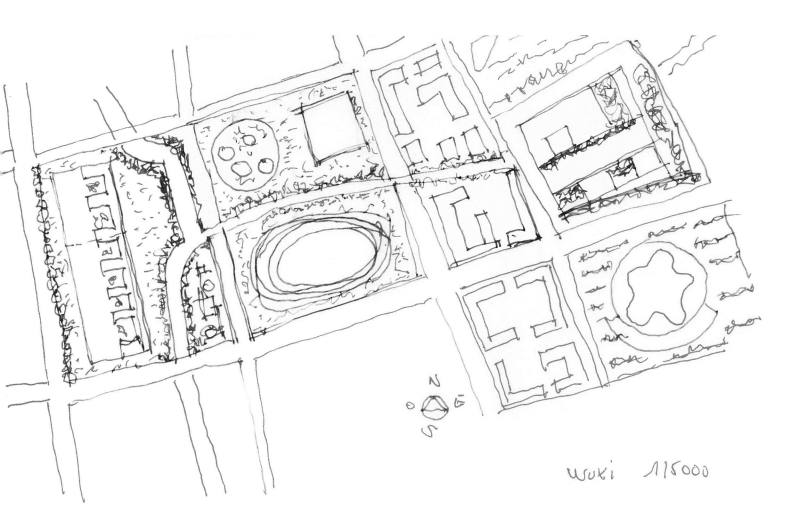

WUXI 1/5000

法国消化系统癌症研究所
IRCAD France

2021年
斯特拉斯堡 / 法国
Strasbourg / France

斯特拉斯堡 / 法国
Strasbourg / France

每年，位于斯特拉斯堡大学医院内的法国消化系统癌症研究所都会接待来自世界各地的5500多名外科医生，对他们进行机器人微创手术技术培训。为此，其大楼已经经历了多次改造。

研究所计划引入10个新的机器人，增设一个礼堂和一个自助餐厅，同时改善原有动线。根据要求，扩建楼需要与原有建筑保持一致，被命名为"法国消化系统癌症研究所3号楼"。

该项目的选址和体量受到了多方面的限制——场地位于原有庭院内，这里曾是一个花园和一个停车场。建筑高度只能为一层，以免影响周围建筑的视野和光照。基于这样的情况，设计团队最终决定向地下建设——将建筑主体"隐藏于"12米的地下。这可以称得上是一项技术壮举，尤其是该建筑距离伊尔河仅有几百米的距离。复杂的施工现场将地下防水与网络和管理设备结合在一起，并通过尖端技术实现。

扩建楼直接从法国消化系统癌症研究所1号楼延伸出来，体量简单与紧凑，并通过玻璃长廊与其相连。一层南侧设有自助餐厅，其一直延伸到室外露台上。玻璃屋顶将整体建筑一分为二，形成了一条东西轴线。

北侧设有礼堂，并一直延伸到地下一层。地下两层包括办公室、技术室以及一个超过300m²的大型实验手术室。

扩建楼沿街展开，成为整个场地内全新的主立面。其外观与其他建筑处理方式相同，以完美地融入周围环境中。外墙装饰着与新民用医院和临近的大学医院相同的灰蓝色铝制覆层，建筑一侧采用法国消化系统癌症研究所2号楼相同的金属网。垂直的烟灰色铝制遮阳装置用于保护玻璃幕墙，这与周围建筑相似。

扩建楼屋顶被视作第五立面，如同花园一般，用于恢复绿色庭院的形象。

室内空间采用木材打造平滑的曲线造型，恰好与建筑外观形成鲜明对比。白色可丽耐板条吊顶蜿蜒起伏，一直延伸到北侧礼堂内。研究所最先进的设备放置在一个超级现代的空间内，以满足多种研究需求。

Every year, the Digestive Cancer Research Institute welcomes more than 5500 surgeons from all over the world to train them in mini-invasive surgery. Located on the grounds of the University Hospital of Strasbourg, the IRCAD building has already undergone several transformations.

With plans to acquire 10 new robots, add an auditorium and a cafeteria, and improve circulation, it became clear that a new extension was needed in continuity to the existing buildings: the IRCAD 3.

Many constraints limited the position and volumetry of the project, which stands in the existing courtyard where a garden and a parking lot used to be. The building is limited in height to one level to avoid obstructing the view from the neighbouring buildings and the light from entering them. The studio decided to build underground and bury the building 12 meters below ground level, a technical feat especially since the building is located a few hundred meters from the Ill River. The complex construction site combines below ground waterproofing with networks and utilities management to equip the basement with cutting-edge technology.

The simple and compact volume of the extension flows directly from the existing IRCAD 1 building and is connected to it by a glazed gallery. On the south side, the first level houses the cafeteria, which extends on a terrace outside. A glass roof splits the building in two, creating an east-west axis that extends into the adjacent building.

To the north of this axis, the new auditorium is set on the ground floor and extends down to the first underground floor. The two underground levels contain offices, technical rooms, and most importantly a large experimental operating theatre of over 300 m2.

The IRCAD 3 lines up along the street and becomes the new main facade of the institution. The exterior features of the extension are treated in the same way as the other IRCAD buildings in order to blend in with the decor. The facades are adorned with the same grey/blue aluminium cladding as the new civil hospital and the neighbouring IHU. The metal mesh used for the IRCAD 2 now covers one side of the new building. Vertical anthracite aluminium shading devices protect the glazed facades, similarly to what was done to the surrounding buildings.

Visible from the peripheral buildings, the roof of the extension is treated as a 5th facade. It is intended to be perceived as a garden to restore the image of a green courtyard.

The interior design contrasts with the building's exterior image by using smooth curves and wood. A dropped ceiling made of undulated white Corian strips runs across the entire ground floor ceiling and extends into the auditorium. The IRCAD's state-of-the-art equipment is housed in a bright ultra-contemporary facility tailored to the various needs of the research institute.

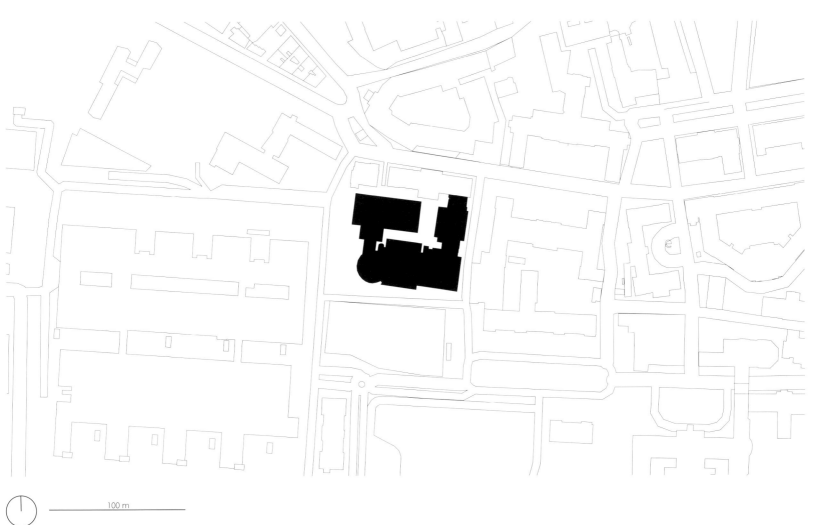

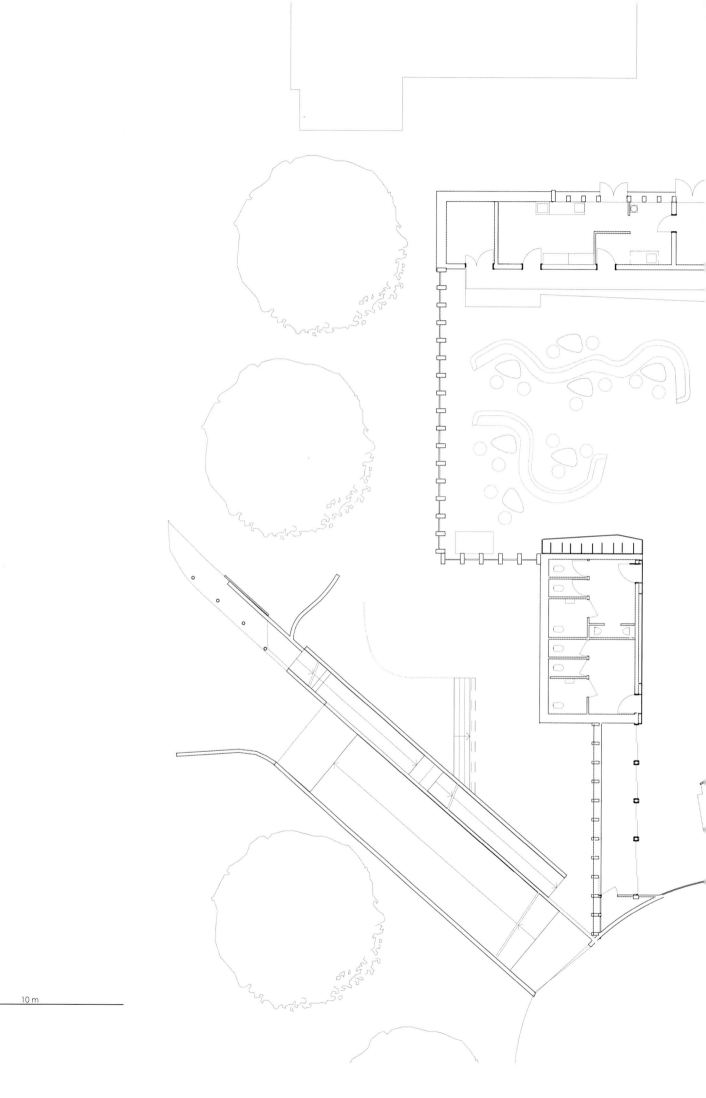

10 m

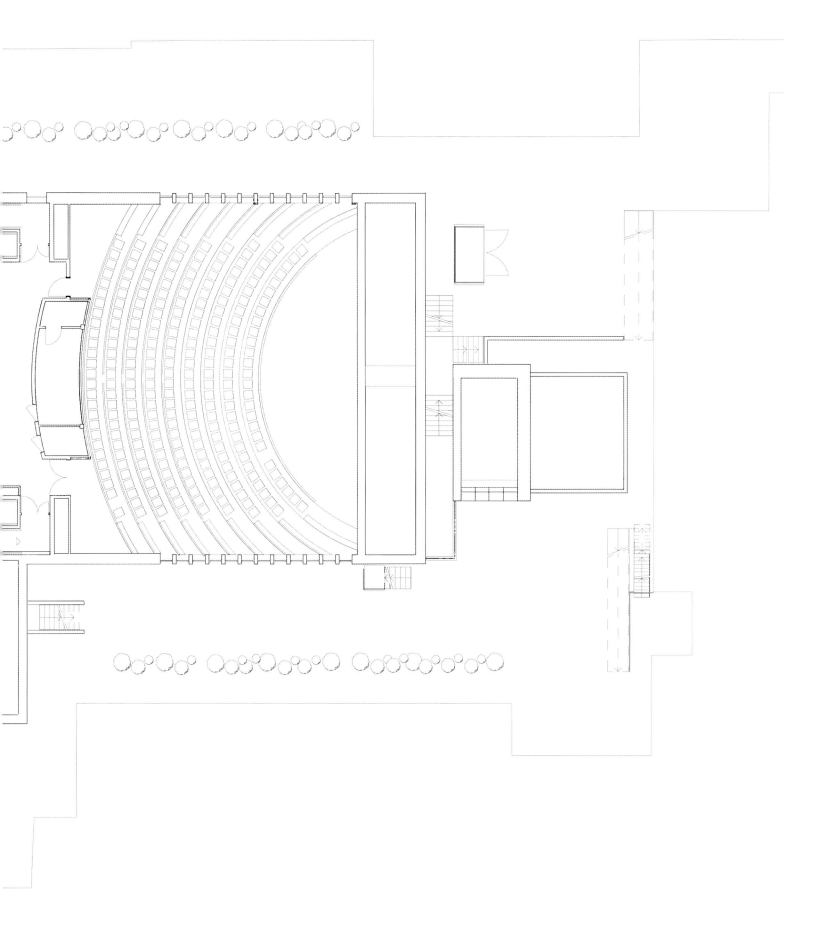

10 m

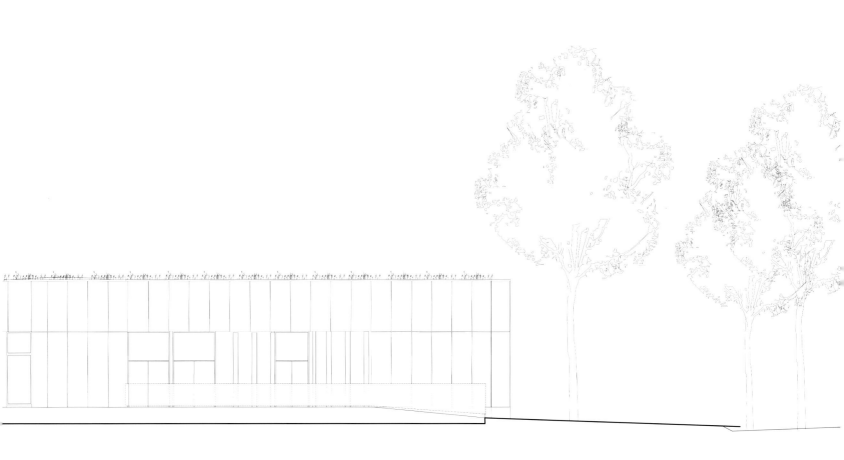

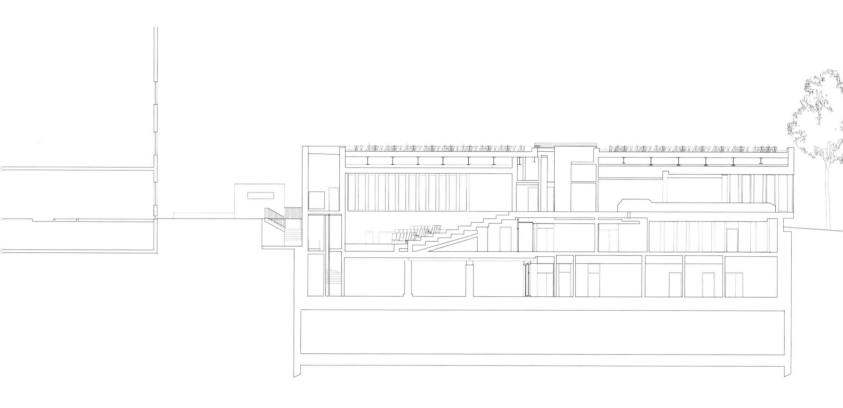

10 m

住宅与商店
Housing and Shops

2021年
斯特拉斯堡 / 法国
Strasbourg / France

新圣于尔班街区作为斯特拉斯堡向克尔街区发展延伸的一部分，获得了诺伊多夫郊区的认可。该项目是莱茵大道沿线的新建筑，在星形广场下方的隧道出口即可清晰看到。

这一地块由两个相邻的项目构成：一个底层设有商店的54个单元住宅楼，一个为经济适用房（专为生活窘迫的人们打造的建筑）。

这两个建筑是分开建造的，但共用一个底座和一个连续屋顶。从外观看来，这着实令人惊叹。

屋顶元素是两个项目的共同主题。新圣于尔班街区内的小型建筑和阿尔萨斯别墅群构成了独特复杂的屋顶景观。而这两座建筑特有的传统大屋顶为整体注入了更多的趣味性。

经济适用房外墙采用清水混凝土预制板打造，营造出坚固结实

的感觉，让人倍感安全。清水混凝土预制板的组合方式以及规整的方形开口赋予建筑更强的韵律感。

住宅楼被分成两个不同尺寸的单元，构成了莱茵大道拐角一处独特的风景。此外，如此的设计便于居民进入位于中央的绿地，也能让居住者在屋内欣赏到不同的景致。

从二层开始，住宅楼外墙全部采用双层表皮打造，增添了一道屏障，使公寓内居民能够免受莱茵大道的噪声干扰，也有助于减少夏日强烈的光线照射。

大面积金属结构更好地衬托了蓝色瓷砖图案，也为居民提供了可以自由使用的长阳台。从外面看去，立面上色彩不一的装饰板宛如在跳舞一般，格外迷人。

As part of Strasbourg's ongoing development towards Kehl, the new Saint-Urbain block takes part in the requalification of the Neudorf neighbourhood's outskirts. Clearly visible from the exit of the tunnel passing under the Place de l'Etoile, the project can be considered as the figurehead of the new constructions along the Avenue du Rhin.
The plot is made up of two adjoining but different projects: the construction of a 54-unit residential building with shops on the ground floor as well as the construction of a home for people in precarious situations.
The two buildings are built separately but are seamlessly linked by a common base and covered by a stunning and continuous roof.
The « roof » concept is the common theme in both projects. The existing urban fabric of the Saint-Urbain neighbourhood is made up of small buildings and an Alsatian villa which form a distinctive and complex roofscape. The two new buildings join in this playful sight thanks to a monumental roof that reclaims the shape of traditional roofs.
The project of the home is built with prefabricated panels of exposed concrete adding a feel of density and sturdiness to it. With these facades the residents feel safe and sheltered

which is appropriate for a building acting as a refuge for people in need. The pattern of concrete panels and the square openings add a regular rhythm to these facades.
The residential project is separated into two buildings of different dimensions. The size of the two units increases crescendo so as to reach an urban dimension and mark the corner along Avenue du Rhin. Separating the project into two parts also provides visibility and access to the green space in the centre of the block. This encourages visitors to discover the varied landscape from the inside.
From the first floor onwards, the facades of the project are coated with a double skin. This second facade forms a screen that shields the flats from the acoustic nuisance of the Avenue du Rhin, preserves the sight of the Saint-Urbain cemetery, and contributes to reducing overheating during the summer.
The large white metallic structure displays a pattern of blue-tinted tiles, and provides the inhabitants with a series of long balconies that they are free to occupy. The facade, designed in depth, is inhabited. From the outside, the uneven rhythm of the panels on the facades confers a dance-like appearance to them.

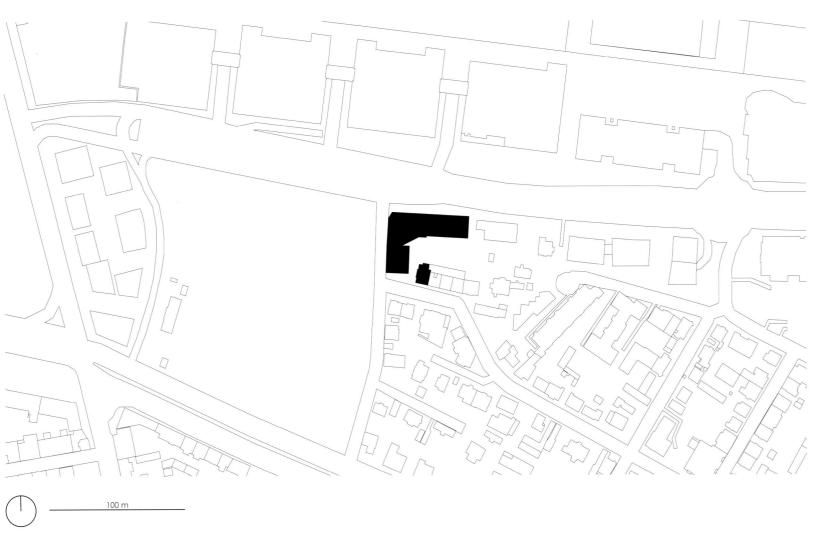

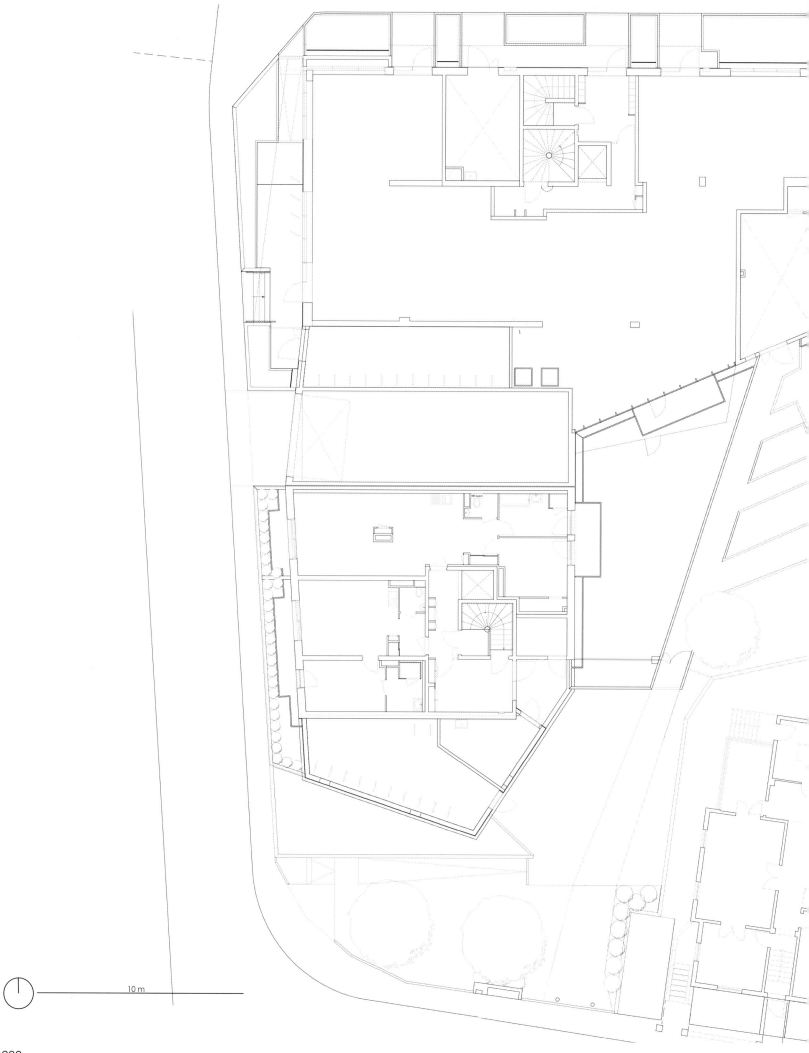

10 m

288

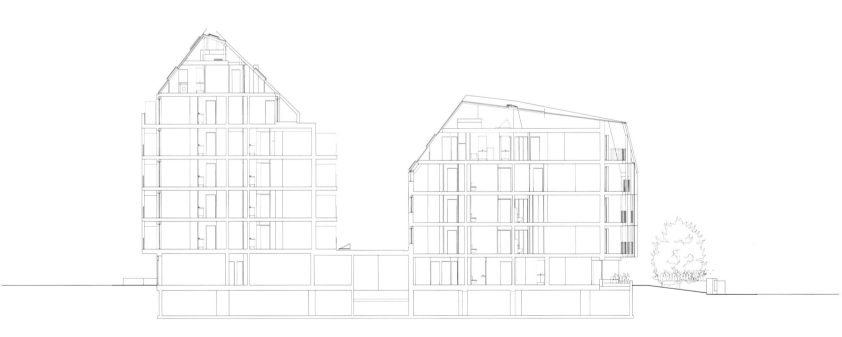

10 m

附录 Index

帕特里克•施韦泽 Patrick Schweitzer
国际建筑作品 2001—2021 Architectural International Projects & Works 2001 – 2021

项目与作品参与者
Participation in projects and works

S&AA事务所及其2001—2021年的全部员工
S&AA and all its employees from 2001 to 2021

参与本书编写
Collaboration on the book
（版式设计、制图、文本编辑、翻译等）
(Layouts, geometric, texts, translations, etc.)
塞巴斯蒂安•布吉翁 Sébastien Bouguyon
艾玛•卡洛 Emma Calò
吉恩•德鲁伊切 Jihène Derouiche
安妮•基尔曼 Anne Kirrmann
尤利娅•玛卡洛娃 Yuliia Makarova
玛丽•奥塔维亚尼 Marie Ottaviani
舒尔德•凡•德•霍克 Sjoerd van der Hoek

294